Ecological Entomology: Insect Life in Odd Environments

Prof. T.N. Ananthakrishnan
and
Prof. K.G. Sivaramakrishnan

SCIENTIFIC PUBLISHERS

Published by
SCIENTIFIC PUBLISHERS (INDIA)

Jodhpur –

5 A, New Pali Road
P.O. Box 91
Jodhpur - 342 001 INDIA

Delhi –

4806/24, Ansari Road
Daryaganj
New Delhi - 110 002 INDIA

Print: 2024

ISBN: 978-81-7233-509-0

Visit the Scientific Publishers (India) website at
http://www.scientificpub.com

Printed in India

The Acrid smog that hangs over
our great cities, the wastes that
foul our rivers, lakes and
coastal waters, oil slicks that
ruin beaches and harbours, all testify
to the urgency of man's need to develop
what has been called
``ecological conscience''

PREFACE

One of the principal aims in writing this book pertains to the increasing impact of human influence on natural habitats in recent years, more notably to the loss of habitat quality. Increased urbanization, considerable reduction in natural forest areas, changing ecology of fresh water ecosystems, lakes and wet lands, besides changing microhabitats, have had their impact on insect populations. A wide range of responses exist among several species of insects living in these changing environments, more especially landscape dynamics creating a shifting mosaic of habitats. Further, loss of habitat quality decreases population growth rate, in course of time resulting in a reduction or extinction of species. Sudden changes in the extent or quality of habitat tend to influence the behaviour, reproduction and survival of individuals. In very rare cases the existence of a particular species depends on the presence of some other species, not to mention of the extinction of food chains. As such, a basic understanding of the nature of habitat loss and fragmentation and their impact on insect species dynamics become important. Fragmentation not only causes loss of amount of habitat, but tends to create small, isolated patches besides changing the properties of the remaining habitat. Some of these aspects are discussed with local examples as far as possible, to provide an overall idea of the need to study environmental aspects of insects in diverse deteriorating as well as odd environments such as fallen logs, deteriorating wood, leaf litter, forest canopy, moss, lichen and fern infesting insects, besides insect ecodynamics in changing agrolandscapes including habitat edges or ecotones. The diversity of aquatic insect fauna in isolated and polluted water bodies as well as some aspects of urban ecology pertaining to changing environment and loss of species, are briefly discussed. Insects occurring in forest canopies for instance, have received very little or practically no attention in the Indian context and their importance is being increasingly recognized. Besides, an understanding of bioindicators reflecting their utility in identifying habitat disturbance, pollution and climate change has also become obligatory for a proper assessment of ecosystem responses to environmental perturbation. This effort is further kindled by the senior author's interest in vistas of insect diversity in natural and interfered forests such as deforestation, to provide for mono-culture plantations such as eucalyptus, teak, tapioca, rubber etc. in many parts of the Western Ghats, such as the Tenmalai area. Similarly the long standing interest of the junior author in the area of biodiversity of aquatic ecosystems as well on the role of indicator species, has made possible the inclusion of relevant information in this area, besides the adaptive diversity of insects to climate, land use and pollutants. Numerous questions are raised about patterns and timing of the evolution of insect-plant association which are essential for an understanding as to ``why the world is the world as it is today".

In this effort several specialists on diverse aspects have generously rendered assistance through providing literature and photographs and while appreciating their generosity, we would like to thank them individually for all the assistance rendered without which the production of this volume would be impossible. The senior author has spent considerable time at the Library of the University of Minnesota, St. Paul Campus, Minneapolis and sincere appreciation is due to Prof. Ralph Holzenthal of the Entomology Department. To Dr. Ananthanarayanan Raman of the Charles Stewart University, Orange, Sydney, Australia, the senior author is indebted for his ready response to requests for a large number of reprints of papers. To Dr B.K. Tyagi of the Centre for research in Medical Entomology (ICMR), Madurai; to Prof. K. Natarajan Emeritus Professor Centre of Advanced studies in Botany, University of Madras, Chennai, to Dr. G.N. Hariharan of the MSSRF, Chennai, to Dr. Sabu Thomas of the Department of Zoology, St. Joseph's College, Devagiri, Calicut, to Dr. R.V. Varma of the Kerala Forest Research Institute, Peechi, Kerala, to Dr. K.A. Subramaniam of the centre of Ecological Sciences, IISC, Bangalore, to Dr. R. Sunderaraj, of the Indian Institute of Forestry Research, Bangalore, to Dr. K. Venkataraman of National Biodiversity Authority, Chennai; to Dr. S. Prasanth Jacob, Dr. K. Murugesan and Dr. K. Sasidharan of the Institute of Tree Breeding and Forest Genetics, Coimbatore, to Dr. G. Thirumalai Joint Director, Zoological Survey of India, Chennai, to Dr. G. Ravi, Rice Research Institute, Aduthurai and to Dr. Alexander Jesudasan of the Madras Christian College, Tambaram our sincere thanks are due for assistance through providing photographs.

LIST OF ILLUSTRATIONS

CONTENTS

CHAPTER 1

ENVIRONMENTAL DETERIORATION AND INSECT DIVERSITY SPECTRUM : AN INTRODUCTION

Insects are one of the most vital components of the earth's biota, being associated with every conceivable type of environment from the most hospitable to the utmost hostile ones. Mass extinction at the end of the Permian Era greatly reduced their abundance. However, their quick recovery and increase in diversity were evident in the Mesozoic, primarily through a complex interdependence with green plants, resulting in extensive adaptive radiation enabling them to occupy all types of plant feeding niches. The evolutionary diversification of the angiosperms is reflected in the diversification of insects. Changes in plant diversity and plant architecture as well as landscape diversity strongly influence the distribution of many species. With deforestation and emergence of new plants, species composition changed from endemic and native taxa to those that have evolved in different regions. The exuberance of insect species in the soil and litter is a further example of this adaptive radiation. The nutritional status of vegetation also affects the abundance and the incidence of herbivorous insects. In natural ecosystems, interactions between insects and plants also become influenced by predatory and parasitic species. Besides, weeds also play an important role since they are indirect mediators of insect-plant interactions facilitating a three-way interaction between weeds, crops and insects. The alarming destruction of tropical forests has sufficiently hindered continued understanding of undisturbed ecosystems. One of the reasons for concentrated biological diversity in the tropics is the closed nutrient cycles supporting highly productive forests on soils with very low fertility, bacteria, fungi and myriad invertebrates participating in the release of nutrients. Mycorrhizal fungi, which are soil organisms forming a network of strands enable capture of soil nutrients, the fungi invading the roots, obtaining nourishment. Tropical deforestation therefore threatens biological diversity.

Human interference with fragile, montane forest ecosystems in general, and tropical rain forests in particular, has caused widespread concerns, regarding the consequences of such interference, which has been known to affect climate, biodiversity and human societies. In short, habitat loss, fragmentation and degradation have been responsible for the disappearance of several species, causing serious concern. Whereas fragmentation creates isolated patches of unmodified habitat (islands), in a highly modified matrix, degradation modifies habitats more or less uniformly over large areas (Lawton, 2000). The preservation and protection of forests to safeguard the natural ecosystems as well as the genetic pool they contain are important aspects, enabling sustenance of food-webs or trophic complexities in not only forest ecosystems, but also others like agroecosystems. Because of their abundance, species richness and importance in the overall functioning of natural ecosystems, they are particularly suited for monitoring landscape changes acting as excellent indicators of biodiversity. Key to successful conservation as such involves an integrated knowledge of biosystematics, ecology, behaviour and genetics.

Insect community responses to habitat constraints and degradation

Charles Elton (1933) has designated the `habitat' as the address and `niche' as its profession, for any living organism. This applies pre-eminently well to both terrestrial and aquatic insects, which are habitat and microhabitat specialists. Dung beetles, tree hole inhabiting mosquitos, bark gleaning insects, shredding caddisflies such as *Anisocentropus, Ganonema and Lepidostoma* with leaf and mineral cases of various shapes and hues are instances in point. Insects in particular, form amazing `species guilds', which are taxonomically different, but ecologically identical species complexes found at great heights in the canopy. Natural and man-made impacts on this environment tend to disrupt their intricate community structure through habitat fragmentation, degradation and niche overlap. However, distinction has to be made between habitat loss and fragmentation in relation to biodiversity, the former having a greater impact than the latter. Impact of fragmentation also depends on the amount of habitat in the landscape besides increase in the number of edges in fragmented habitats, which tend to cause changes in the to and fro movements of the individuals. Survival of species depends on the size of the surviving patches and species occupying higher trophic levels tend to be more vulnerable to habitat fragmentation than those in the lower food chain (Fahrig, 2003). Selectively logged and plantation forests are degraded forests, with fragmentation and degradation generally operating together. Degraded forests lose species with varied responses, such as those in the canopy (Marshall and Swami, 1992). Fallen tree branches, roots and dead leaf surfaces offer favourable habitats for the survival of insects.

Arthropod populations are very sensitive to environmental impacts including fragmentation, disturbance, habitat modification, ecological disruption, climate change and chemical pollution. In response to decreasing patch size, species abundance and richness declined. Selected arthropods evince strong responses to habitat fragmentation, acting as effective indicators, providing early information on ecological consequences of fragmentation. Relevant examples pertain to the decline of dung composition due to decline in the abundance of butterfly species. As such, different species use different kinds of habitat, not to mention of many species requiring more than one kind of habitat within a life cycle (Fahrig, 2003). Besides, habitat loss tends to reduce trophic chain length, alters speceis interactions, reduces number of specialist species and affects breeding and dispersal success (Fahrig, 2003). It is obvious that selected arthropods show strong responses to habitat fragmentation, acting as effective indicators providing early signs of ecological consequences of fragmentation. Landscape ecological studies therefore focus on habitat fragmentation and its influence on distribution, movement and persistence of species. In short, landscape is a mosaic of habitat patches and interconnections (Turner, 1989). The shape, size and diversity of patches influence patterns of species abundance.

As regards functional responses to fragmentation, it is their role in ecosystem processes that is important, notably decomposition and nutrient cycling. The major functional groups pertain to the pollinators, herbivores, natural enemies and decomposers. For example, a reduced population of pollinators and their diversity tends to result not only in reduced population, but also in reduced seed set, lower reproductive success and reduced genetic viability in small fragments of forests. Interestingly enough increased abundance and biomass, typical of the edges of fragmented areas in forests, tend to become reduced and one of the good examples relates to changes in beetle species components in small fragments. As such, the role of forest edges in providing habitats for several arthropod species has come to be recognized, representing the `edge effect' (Didham, 1997).

Forest Edges

Places where plant communities meet or where successional stages within plant communities come together are referred to as forest edges, whose environment significantly differs from the interior of adjacent patches. Such differences relate to biomass, soil characters, species composition and abundance and hence are regions of continual change (Foreman, 1986). Four kinds of forest edges are known - permanent (forests next to lakes), permanent anthropogenic (forests next to agricultural or urban areas), natural successional (edges caused by fire or winds) and anthropogenic successional (edges caused by timber harvesting). Both microclimatic attributes and ecological processes are also involved and because edges abut two quite different habitats rich in food and cover, ecological studies relate to species abundance and richness near edge habitats (Kremster and Bunnell, 1999).

Habitat fragmentation has been perceived as the principal threat to most species or the single greatest threat to biological diversity. Habitat fragmentation is mostly caused by the expansion and intensification of human land use. Fragmentation of habitats is characterized by three important processes:

a. area reduction of the original habitat in the landscape due to habitat loss;

b. area reduction of the emerging habitat fragments and

c. increasing distance between fragments.

As a consequence of the reduced land area, edge effects may have additional impact on diversity and distribution pattern of species (Kruess and Tscharntke, 2000). In modern agriculture, habitat loss on a landscape scale often reaches 80% or more and at such higher levels of fragmentation, isolation appears to be a major threat to biodiversity. Further, the effects of fragmentation of a particular species depend on its ecological requirements or biological characters. Polyphagous species are less sensitive to fragmentation processes than monopha-gous ones. The principal processes shaping both the communities within forest fragments are area effects, edge effects, shape of fragments, degree of spatiotemporal isolation and degree of habitat connectivity in the landscape (Watt, Stork and Hunter, 1997). Fragmentation therefore results in a reduction of habitat area followed by a decrease in living space for animals and plants tending to result in reduction of species richness. In short, fragmentation of habitats negatively affects species diversity, abundance and interaction, since area loss and habitat isolation dramatically reduce species diversity. Pollinating insects are greatly affected by forest fragmentation not to mention of termites which show a significant impact with lower species numbers and rarer species. In short, increase in fragmentation tends to affect insect population by reducing habitats, including the management practices for field boundaries, so essential for the sustenance of many species.

``Survival of an individual or population within a given landscape depends on the integrity of the functional units measured as connectivity and heterogeneity of the landscape mosaic" (Ekbom *et al.,* 2000). Connectivity between landscapes relates to the degree to which local populations are joined or connected and their influence on the population dynamics of species. Connectivity is the extent to which an organism is able to move through the landscape to access resources vital to its survival and this is an important component in landscape structure, or is a functional link among habitats. A good example relates to the damselflies which could persist in mixed landscapes by flying across open pasture, linking their resource patches together (Taylor, 2000).

Fragmentation results in a reduction of habitat area and consequent decrease in living space for plants and animals, reducing species richness. Forest fragments are complex habitat islands, with butterfly species in general, showing a significant decrease in species richness. Pollinating insects are greatly affected by fragmentation. Termites, like ants show a significant species-area relationship in forest fragments, with small fragments having lower species richness and more rare species. Functional groups such as pollinators, herbivores, predators/parasitoids and decomposers are adversely affected by forest fragmentation. A two-fold decrease in species richness and a seven-fold decrease in average population of dung beetles, result in dramatic reduction of the rate of dung decomposition (Watts, Stork and Hunter, 1997). The number of species lost depends on where the habitat is removed, because species ranges rather than their abundance determine which species is lost (Turner, 1989). Long term stability of the landscape depends on climate, biota and soil. Landscape ecological studies relate to the effects of spatial patterning and changes in landscape structure such as habitat fragmentation and impact on distribution, movement and persistance of species. Wilcove *et al.* (1986) defined habitat fragmentation as ``a large expanse of habitat which is transformed with a number of smaller patches of smaller total area isolated from each other by a matrix of habitats unlike the original".

Hanski (2005) has indicated four kinds of habitat loss viz., i. loss of quality, resulting in reduced population growth rate, increase in population fluctuation and reduced population size and increased risk of population extinction; ii. loss of quantity (area); iii. reduced emigration and colonization rates and iv. loss of continuity (reduction in emigration rate habitat occupancy). Studies of microhabitats, in particular, become important and relevant in terms of changing habitats as a result of fragmentation and discontinuity. Such microhabitats could be observed as dung heaps, corpses of dead animals, decaying remains of plants and small bodies of water. Microhabitats are habitats for small-bodied species and they are very significant for overall species richness.

Many aquatic insects have evolved adaptations as related to habitat constraints (Danks and Rosenberg, 1987). The adaptations of insects in wetlands may be a response to habitat constraints such as fluctuating supplies of moisture and acidity and consequent low nutrient availability. Some marsh inhabiting insects share features of insects inhabiting temporary pools, where conditions are unpredictable and food resources very scare. For example, small size, precise microhabitat selection, tolerance to environmental extremes and extremely short life cycles (baetid and caenid mayflies in the tropics) are a few notable adaptations. Multivoltinism and parthenogenesis are further adaptations of some aquatic insects due to unpredictable conditions of their habitats.

Poor urban planning in relation to sewage treatment has adversely affected habitat association of several species of butterflies (swallowtails) sensitive to land use and highly polluted environment. Human induced changes in environments, involving habitat loss and fragmentation have become so conspicuous that their ecological consequences result in considerable deterioration in quality of the environment. As such, in recent years, there is a shifting of research interest more especially in landscape ecology which deals not only with major habitat types, but also with other microhabitats. Apart from factors influencing risk of extinction of local species in individual habitat fragments, migration among fragments is an important process for the establishment of new species.

Conclusion

The unanswered question relates to how far forests are affected by fragmentation and loss of biodiversity. Are arthropods important in ecosystem function and does fragmentation alter trophic interactions and energy flow in the canopy? One cannot overlook the fact that habitat loss will soon outstrip global climatic change! Such multilevel interactions need to be identified in fragmented forests. Species tolerance to different degrees of deforestation and fragmentation and the size of species ranges are aspects which need attention.

CHAPTER - 2

ADAPTIVE DIVERSITY OF INSECTS TO ENVIRONMENTAL CHANGE

Introduction

Throughout the history of the earth, climate has changed continously with major effects on the structure of insect communities. Periodic Advances and retreat of continental glaciers in the recent geological past from the Pleistocene have been modifying the diversity and distribution range of insects, besides creating many unique landscapes and habitats. In his classic paper on `Insects and Climate', Boris Uvarov (1931) discussed the problem of predicting the response of insect populations to environmental changes, indicating that ``the climate of every area on the surface of the earth has undergone a series of great changes throughout geological history. But the difference in the changing climatic patterns in recent years have been augmented at an unusually fast rate, thanks to the anthropogenic environmental changes mostly degrdation or destruction of the world's natural habitats, as well as the combined effects of habitat loss, fragmentation, pollution and climate changes which have posed a serious threat to the world's biodiversity. Since the geographic distribution of species is a consequence of their population attributes, the size of a species range is naturally related to the local abundance in the favourable parts of the habitats. The capacity to tolerate various climatic conditions during unfavourable seasons is a characteristic of several species of insects. While individual species respond independently to climate, the sum total of individual responses makes up the response of the communities.

Impact of climate change

Climates have usually changed slowly in the past, but in today's scenario, rapid changes in climate are evident through increasing concentration of CO_2 and other greenhouse gases in the atmosphere. Over the last century, the amount of CO_2 in the atmosphere has increased by more than 2.5% as a result of deforestation and combustion of fossil fuels. Global temperature is likely to increase by a few degrees over the next five to ten years. Such warming would trigger insect movement and establishment at high altitude and latitudes. Limited land area with elevational increase and greater warming at higher latitudes tend to hinder the normal range of insects. Both gaseous pollutants and increased CO_2 concentrations are likely to alter the amount of insect damage to trees and this increase tends to be through larger populations because in such environments, herbivorous insects cannot be controlled by natural enemies. Further, in elevated CO_2, chewing insects show greater consumption rates, to compensate for poor food quality. Elevated atmospheric CO_2 tends to have considerable effect on phytophagous insects ``by interfering with CO_2 gradients used by adult females to detect and evaluate their hosts (Strange, 1977), although indirect effects via changes induced in plants by CO_2 exist". Some species increase feeding rates at higher levels of CO_2 and the responses tend to be species specific. There is a need for consistent and wide range information related to the impact of high CO_2 levels during the developmental periods of early instars of Lepidoptera. CO_2-induced changes tend to have wider implications on soil microarthropods through their potential to alter soil structure (Rilling *et al.,* 1999). Collembolan populations respond to elevated CO_2 and with the increase of soil fungi on which they feed, resulting in the shifting of species abundance, as for instance, *Folsomia* replacing *Proisotoma* in elevated CO_2 (Bezerner *et al.,* 1998). As such, the multidimensional effects of gaseous air pollutants on herbivorous insects especially in developing countries with rapidly expanding urbanization, industrialization and increased energy consumption need investigation.

Thus, changes occur at all levels, in total richness and species composition. In short, the end result in the most degraded habitat is a drastic reduction in species richness. Food webs tend to change dramatically. ``Species-area relationships are not universal, but they are one of the oldest and most regular patterns in ecology" (Lawton, 2000). Species occupying higher trophic levels should be more vulnerable to habitat fragmentation than those feeding in the lower food chain. In this context, the need to document the significance of extreme climatic events as well as non-climatic variables limiting distributions, becomes necessary. The use of geographic information systems and other data from a variety of sources including satellite imagery and high resolution soil, host plant, and animal maps is vital. Studies should also be made to explore whether the insect species permanently occupy new and favourable local habitats colonized during extreme climatic events, and how frequently do they adapt genetically to the new environments (Sutherst *et al.,* 1995). Ecotones, which are boundaries between major types of vegetation, may particularly be sensitive to changes in climate. Insect populations dependent on particular types of vegetation may change rapidly in abundance and distribution at vegetation ecotones due to climate change, because their habitats are reorganized around them. As such, rapid changes in the distribution and abundance of many species tend to result in decreased future communities due to failure to migrate across increasingly fragmented landscapes as well as increase in globally rising extinction rates. Community-level responses to global change, particularly climate warming will lead to shifts in the geographic ranges of many species, gradually destroying existing combinations of species and creating new ones. Accelerated pace of changes in altitudinal and latitudinal species distributions will be the predicted pattern in future (Lawton, 2000).

Impact of land-use changes

Forest conversion to rubber plantations in Kerala and expansion of coffee plantations in Coorg and large scale deforestation in the Himalaya and other montane regions have resulted in secondary formations with associated changes in floral and faunal composition. In many places, population pressure is responsible for deforestation and land degradation. Deforestation has ecological effects viz., habitat loss, fragmentation and edge effect resulting in the loss of biodiversity. Decline in the biological productivity of an ecosystem will lead to desertification soil degradation, alteration in available water supply and yield decline. Secondary bamboo forests have declined in favour of paper industry, resulting in the loss of several bamboo infesting species. Plantation forestry such as teak, sal and eucalyptus has changed the faunal scenario. The rich diversity of traditional agro-ecosystems is being lost, including crop diversity at the specific and subspecific levels. Modern agriculture is based on the use of high yielding varieties replacing biodiversity in agriculture. The large size and uniformity of crop monocultures tend to fragment the natural landscape affecting diversity and abundance of natural enemies, since larger the area under monoculture, lower the viability of a given population of natural enemies. Such landscape changes have led to increased insect outbreaks. Efforts to protect habitats must also be guided by monitoring indicator species such as Hymenoptera. Their effective population size and viability are to be maintained. It is important to recognize subspecific, species specific and functionl groups level biodiversity, in relation to ecosystem functioning.

Fig. 1. Fragmentation of landscape: natural and fragmented forests

Fig. 2. Teak skeletoniser and teak defoliator. (After Prasanth Jacob) (Top) *Hyblaea peura* (Bottom) *Eutectona macharalis*

The impact of land use changes on insect communities driven primarily by ecological, socioeconomic and political factors may be gauged to some extent when we study such impacts in one of the two Indian hotspots of biodiversity viz., the Western Ghats with its long ecological history dating from the detachment and northward drift of Indian plate from southern Gondwanaland around 195 million years ago during the early Jurassic. It has also witnessed myriads of land-use changes in the course of one of the longest periods of waves of human settlement. Shifting cultivation in the remote historical past to monocultural plantations and several industrial and damming activities in the recent times have been causing considerable deterioration of its environment quality, resulting in depleted native insect diversity and proliferated introduced diversity. Some of the beetles and bugs that have established as pests of the introduced plants have also secondarily established as pests of a number of indigenous plants and are spreading to other cultivars. The heteropteran, *Antestia cruciata* is an instance in point. More examples can be seen in the outbreak of teak skeletoniser (*Eutectona machaeralis*) and teak defoliator (*Hyblaea purea*) in almost all the teak growing areas of Wesern Ghats and the outbreak of *Pteroma plagiophleps* (Lepidoptera) on *Albizia falcataria.* Introduction of the exotic fodder plant, *Leucaena leucocephala* (subabul) has also brought along its pests, *Heteropsylla cubana* (Homoptera) and *Ithome lassula* (Lepidoptera) to the Western Ghats (Fig. 1). Malaysian creepers as cover crops in rubber plantations are responsbile for the occurrence of the leaf miner, *Japanagromyza indica* in the Western Ghats (Lakshminarayana *et al.,* 2002).

Likewise, studies on the impact of riparian land-use on stream insect communities at Kudremukh National Park located within the Western Ghats of India showed that diversity and community composition of stream insects varied across streams with different riparian land-use types, reflecting fundamental alterations in the functioning of stream ecosystems. Here, the aquatic insect family and generic richness was higher in the streams with natural riparian vegetation than with the human modified ones, in the investigated segments of landscape. The streams with riparian evergreen forests in Kudremukh area are mostly shaded first order streams. On the other hand, the streams with riparian arecanut plantations are partially shaded second order streams. The partially shaded streams had high diversity representing transition of heterotrophic-autotrophic nutrititonal zone of the lotic ecosystem, permitting co-existence of insect taxa evolved in both the systems. The turnover of aquatic insect genera in the streams with different riparian land-use types shows that taxa composition changes with the riparian land-use. Taxa such as *Hydropsyche* (Hydropsychidae), *Macronema* (Hydropsychidae), *Baetis* (Baetidae), and *Isca* (Leptophlebiidae), with tolerance to disturbance characterize streams flowing through human influenced riparian land-use types. Taxa with high sensitivity to human disturbance such as *Helicopsyche* (Helicopsychidae), *Dineutus* (Gyrinidae), *Enithares* (Notonectidae), *Limnogonus* (Gerridae), *Goera* (Goeridae) *and Notophlebia* (Leptophlebiidae) are present in the streams in natural riparian evergreen and semievergreen forests (Subramanian *et al.,* 2005) (Fig. 3). Similarly, in the Cauvery river system, the presence of pollution tolerant or intolerant taxa reflected the riparian land-use patterns, subjected to human impact such as riparian deforestation canopy, decreased shading and increased sediment and organic pollution (Sivaramakrishnan *et al.,* 1995). In the streams with human modified riparian land-use types, proportional abundance of the functional feeding groups viz., collectors and scrapers such as *Hydropsyche* (Hydropsychidae), *Macronema* (Hydropsychidae), *Baetis* (Baetidae), *Isca* (Leptophlebiidae), and *Choroterpes* (Leptophle-biidae) tends to be higher indicating autochthonous food sources of those streams (Subramanian *et al.,* 2005) (Fig. 3). However, it has been argued that in tropical Asia, it is difficult to distinguish changes due to human impact from changes resulting from natural variability at various spatial and temporal scales (Dudgeon, 1999). Interestingly enough, at the edge of Aswan Reservoir in Dehra Dun valley, fully submerged Typha plants, producing a mucilaginous secretion harbour an aquatic thrips, *Organothrips indicus* (Bhatti *et al.,* 1998).

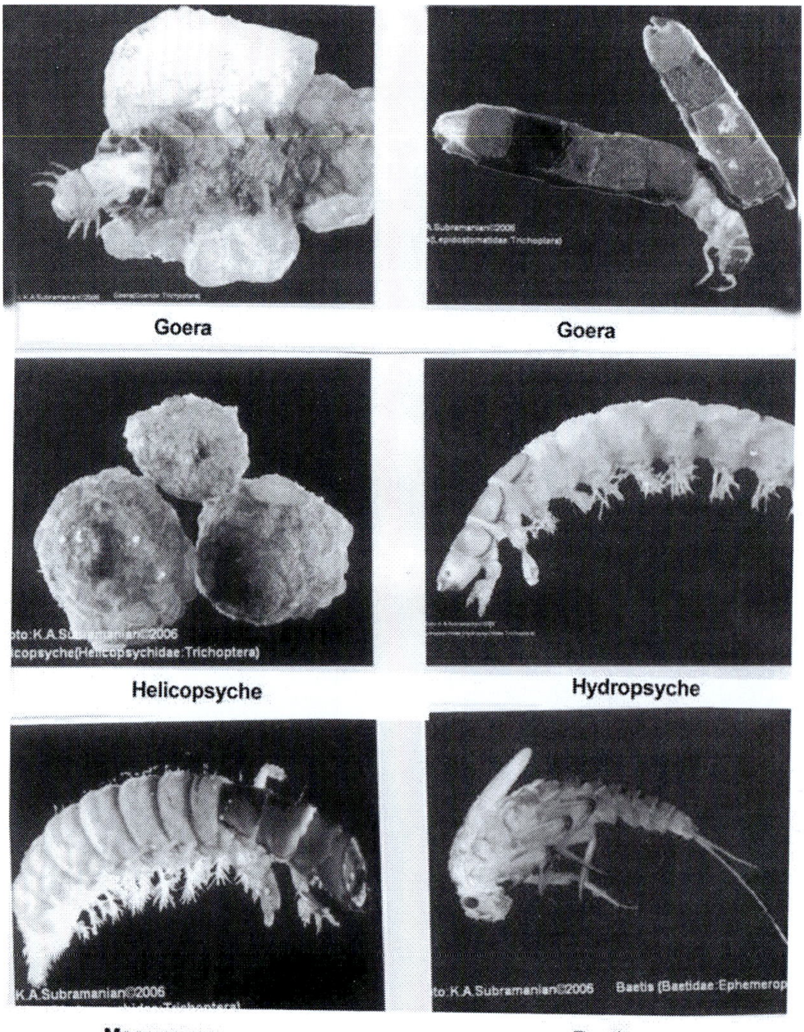

Fig. 3. Indicators of aquatic pollution

Impact of environmental degradation on insect vectors of diseases

Changes in climate particularly in terms of increase in global temperature along with human induced environmental changes at the local scale, including urban agglomeration, deforestation, agricultural and water resource developments and poor sanitation and hygeine due to human population explosion are a threat, in terms of vector-borne diseases as a consequence of range extension, range shift and dramatic resurgence of insect vector and pathogens. Dengue fever, a human arboviral disease has the dubious distinction of spreading at an alarming rate due to its mosquito vector, viz., *Aedes aegyptii*, adapted to breed in man-made containers in tropical urban areas, with eggs resistant to desiccation. Its temperate counterpart, *Aedes albopictus*, extends its range into higher altitudes and higher latitudes with ease as a result of global warming, in addition to low-level viral transmission or transovarian transmission through the overwintering stages of the mosquito. Another arboviral disease, Japanese Encephalitis (JE), transmitted by *Culex vishnui, Culex pseudovishnui* and *Tritaeniorhynchus* (Fig. 4) sp., the rice-field breeding mosquitoes form a great threat in tropical South India. This has a short and well-defined season of epidemic transmission with peak in vector abundance during rainy season. Birds and pigs act as maintenance and amplifying hosts. Filariasis (elephantiasis) is caused by a mosquito vector transmitted disease in which continued urbanization plays equal role as global warming, in most tropical cities since the urban vector, *Culex quinquefasciatus*, has the ability to breed in highly polluted water. Likewise, human induced environmental changes, mainly water resource development, deforestation and urbanization have the most profound effects on malaria.

The effects of these are complex and dependent on the breeding sites and habitat preference of the local vector. Whereas anopheline mosquito vectors usually cannot breed in containers or in highly polluted water, *Anopheles stephensi*, one of the most important vectors of the Indian subcontinent has adapted to breeding in overhead tanks, a strategy for survival and transmission of malaria in towns as well as in the countryside (Lines, 1995). Thus, global warming and anthropogenic environmental changes are both significant with regard to limits of distribution of vectors and of the diseases they transmit, though the former appears to be more significant than the latter.

Conclusion

A vast range of permutation of climatic conditions has to be taken into consideration to be able to predict the distribution and geographic of a number of species of agricultural, forestry and public health importance, in the wake of the emergence of new climatic regimes. Some species are unlikely to survive climatic changes partly due to the loss of genetic variability. Synergistic effects of climate change with habitat destruction have been implied as among the greatest cause of extinction. While climate change could have serious implications for regional fauna and flora, with insects being the first to respond to change, loss of keystone species such as pollinators and parasitoids may have critical effects upon the whole biotic community. Global warming and increasingly fragmented and degraded landscapes are not likely to allow insects and their host plants the opportunity to move up in elevation across latitudes. Insects and climate change remains an active area for further research predicting what tomorrow's pest species are likely to be. An aspect of immediate concern also relates to the impact of human induced climatic changes on the frequency of major environmental disasters, in particular cyclones and hurricanes. Finally, the views of Lawton (2000) cannot be overemphasized, in that ``global change, be it habital degradation or fragmentation or climate change or rising atmospheric $CO_2$ concentration, simply speeds up the process of changing interspecific interactions in time and space". Thus ``with increasing rapidity, the pool itself will start to change, as species become locally extinct and new ones invade, tearing apart existing webs of interaction and creating new ones".

CHAPTER 3

INSECTS AS INDICATORS OF ENVIRONMENTAL QUALITY

Introduction

Community ecological studies have provided sufficient evidence for the disturbance hypothesis which relates to levels of disturbance, providing information on species rarity or loss. This hypothesis is used to assess the health of natural ecosystems and species diversity in disturbed ecosystems such as changed land use patterns, overharvest of natural resources, impact of exotic species, forest fires and soil and water pollutants, which have considerable negative role on natural ecosystems. In recent years, the study of living organisms reflecting environmental disturbances provide clue to organisms which indicate such environmental disturbances. These are termed `*bioindicator species.'* Indicator assemblages should exhibit varying sensitivity to environmental perturbations. Such indicator assemblages or species respond readily to environmental changes which could early be measured. They could be chosen *taxonomically* through monitoring the relative abundance of all members of a taxonomic group or groups of *functionally* representing different ecological roles in their habitat (Kremen *et al.,* 1993). Keystone species are often considered as species, which are important in supporting others, elevating, biodiversity. To increase protection of biodiversity, the utilization of diverse biotic elements in the assessment of land areas for their conservation value becomes important, enabling identification of communities, habitats, ecotones, biotopes and areas of endemism (Kremen, 1994). In this effort, a concept was initiated called `target taxon assemblages', which refer to biogeographically informative assemblages frequently occurring among terrestrial arthropods, which can be chosen taxonomically or ecologically. Many such species are sensitive to environmental impacts including fragmentation, disturbance, habitat modification, ecological disruption, climatic changes and chemical pollution (Kremen *et al.,* 1993). Some terrestrial arthropods like dung beetles, carrion beetles and forest restricted butterflies tend to show strong responses to habitat fragmentation. They can also indicate ecological functioning such as disruption of insect pollinator services due to fragmentation (Jennersten, 1998). Dung decomposition in small, isolated patches of tropical moist forests declined with lowered diversity and abundance of species of dung beetles, which promote dispersal of seeds of a wide variety of plants. They are termed *mobile-link species,* which help in studies of keystone mutualist plants (Terborgh, 1986, 1988). The mobile link species also include the well studied euglossine bees which can link plant species, obligate pollinators, tending also to be critical to the reproductive success of plants (Gilbert, 1980). As such, terrestrial arthropods offer very useful characters as indicator groups for monitoring programmes. The unparalleled diversity of arthropods providing a rich data source and their capacity to rapidly respond to environmental change makes them potential early warning indicators of environmental change (Kremen *et al.,* 1993).

Bioindicators

Traditionally, *bioindicators* have been used to assess ecosystem responses to environmental perturbation, often associated with human land use (Noss, 1990). Ant species richness and composition have indicators of restoration success. In countries like Australia, ants have been incorporated in monitoring programmes associated with fire, grazing and logging practices (Neuman, 1992; Van der Woulde *et al.,* 1997). In particular the use of functional groups in relation to environmental stress or disturbance (Andersen, 1995, 1997). Ant community ecology involving patterns and processes have therefore become relevant. Likewise, termite abundance, biomass and species richness are generally reduced when forests are cleared. Chironomids such as *Kiefferulus, Cryptochironomus, Polypedium* etc. are known to be indicators of water quality, as also several mayflies, stoneflies and caddisflies. Chironomids are indicators of acidification, salinity and heavy metal pollution. Midge larvae with their rapid generation times, quickly respond to environmental changes, as they are able to tolerate low oxygen content due to haemoglobin in their body fluids. They are used to construct a classification of the trophic status of lakes (Thienemann, 1954).

Bioindicators also exhibit particular symptoms or responses indicative of changes in some environments (Hawksworth and Ritchie, 1993). They are successfully used in identifying habitat disturbance, pollution and climatic change. Mycorrhizal fungi indicate forest disturbances by the frequency of fruiting bodies. Changes in forest canopy are identified by ants and Collembola. Earthworms and scarab beetles are fine indicators of soil health. Investigations of the soil-litter inhabiting arthropod fauna of monoculture and polyculture plantations in terms of litter decomposition and nutrient cycling and prey-predator interactions provide an idea of species occurring in relatively fertile ecosystems. Such information also provides the biodiversity baseline for these forest ecosystems which could be used for measuring future changes due to various anthropogenic activities. Freeliving nematodes are also used in soil and water pollution studies. Good indicators of environmental pollution are Cyanobacteria, various algae and diatoms. Many algae are good indicators of water pollution such as Myxophyceae and Chlorophyceae, diatoms and euglenoids. Local waste water contains *Acromonas* and *Klebiella.* Presence of pathogenicity is assessed by counting *Pseudomonas, Streptococci, Clostridium* etc. Bacteria and fungi are also used as indicators of soil quality. Selection of indicator assemblages which tend to respond in different ways to the same stress improves the sensitivity of monitoring (Karr, 1991). What is needed is a study of the changing patterns of bioindicators.

Insect *flagships* are charismatic and are of importance in conservation biology and such flagships have value in detecting to what extent the landscape is truly fragmented. Flagship species may be indicators for the protection of other species. Insects have been used as monitors indicative of the quality of the ecosystems. Quite some experience exists using aquatic arthropod communities in the larger context of bentic macroinvertebrate assemblages for monitoring of water quality. Plecoptera, Ephemeroptera, Odonata, aquatic Heteroptera and Coleoptera have extreme sensitivity to low levels of pollution. As such, indicator species refer to species or species groups that respond readily to environmental stress or change in ways that can easily be measured or observed. Indicator assemblages for monitoring purposes must be sensitive to anthropogenic disturbance and be able to provide an early warning of ecological change. Many terrestrial arthropod populations are sensitive to environmental impacts including fragmentation, disturbance, habitat modification, ecological disruption, climatic changes and chemical pollution. Selected arthropods like dung beetles or butterflies could demonstrate strong responses to fragmentation making them effective indicators providing early warnings of fragmentation. Habitat destruction and fragmentation influence not only the distribution and abundance of terrestrial arthropods, but also their ecological functions expressed in terms of truncated food-web relationships.

Indicator types

The ecological health of certain microhabitats may be monitored by using high species assemblages of terrestrial arthropods such as oribatid mites, millipedes, Collembola, crickets, flies, wasps etc. large-scale diversity of arthropods provide a rich data source enabling identification of rapid responses to environmental changes making them early warning indicators of environmental change. An overall assessment of such indicator species/groups has led to the recognition of different categories of indicator species, such as *sentinels*, (introduced with atypical condition as early warning devices); *detectors* (which show measurable response to environmental change); *explorers* (presence indicating probability of disturbance or pollution); *accumulators* (accumulating chemicals in them) and *bioassay organisms* (detecting presence or concentration of pollution).

In the context of biomonitoring of freshwater bodies, both lotic and lentic, through benthic macroinvertebrates of which aquatic insects form a dominant component, the concept of indicator species is of central importance. Indicator species or species assemblages have narrow and specific environmental tolerances and their presence is a reflection of its environment. Their properties and responses include : biochemical, to life history changes and bioaccumulation at the organism level; and presence/absence or numerical predominance of indicator organisms, populations and species assemblages (Johnson, *et al.,* 1993) Taxonomic soundness and easy recognition by the nonspecialist, cosmopolitan distribution, numerical abundance, low genetic and ecological variability, large body size, limited mobility and relatively long life history, are well known ecological attributes and suitability for use in laboratory studies are considered the essential criteria for an ``ideal'' indicator (Hellawell, 1986). Besides their role as sensitive indicators of water quality due to their responsiveness, directly or indirectly, to changes in water quality and the long-term persistence of useful taxonomic characters associated with the head capsule and mouthparts, Walker *et al.*, 1991 opine that chironomids are excellent indicators of past climate and have developed a model to infer historic temperatures from the remains of Chironomidae in unstratified lakes.

Apart from indicator species, conservation of keystone species is also essential for the well being of any ecosystem. This is because they play a vital role in an ecosystem and upon which a large part of a community depends. Keystone species are also necessary to help reestablish and sustain ecosystem strength and stability. One of the outstanding instances of keystone species relates to epiphytes in forest canopies, which contribute to ecosystem processes such as nutrient cycling, so that forest canopies have become important in maintaining the diversity, resiliency and functioning of forests. The arboreal, litter and soil forest communities support beetles, oligochaetes and isopods and the dominant microfauna include mites, collembola, and ants. Insects provide varied sensitive indication of changes at different tropic levels and hence are suited for monitoring landscapes change in view of their abundance and species richness. The need for interacting with a broad spectrum of ecological niches and microhabitats cannot be overemphasized. Ants, termites, Diptera and parasitic Hymenoptera have been known to act as useful indicators of negative and positive factors of environmental conditions. Cicindelids have been identified as positive indicators or fast indicators of biotope quality and have been reported as very sensitive indicators even at lower scales. Another interesting example relates to several orthopterans acting as good biotope indicators, since their recognition in the canopy has been made possible by their sounds.

Exotic introductions and biological invasions - a threat to environment quality

Apart from localized human disturbances and changing global and regional atmospheric conditions, the threat posed by accidentally and deliberately introduced organisms is real and increasing all over the world disrupting the subtle ecological balance between the local communities already existing in any ecosystem. For instance, major destroyers of the Hawaiian biota are the African big-headed ant (*Pheidole megacephala*). The ant lives in borderless supercolonis of upto millions of workers and breeding queens. These supercolonies eat or drive out a large part of the other insects on their path. Hawaii, because of its extreme isolation had no native ants before the arrival of humans. As a consequence, Hawaii's plant and animal communities evolved to conform to an ant-free world. When keystone species are lost or removed, there is a noticeable effect on the system, often referred to as the cascade effect. Four categories of keystones are recognized, *viz.,* Keystone predators and parasites such as ants, keystone mutualists as in fig wasps; keystone resource species as in ants and keystone dominants (La Salle and Gauld, 1993). While keystone predators are important for the maintenance of diversity keystone matualists are specialized pollination system as in orchids. Bees are the most important keystone matualists absolutely essential for the maintenance of diversity of flowering plants. Further, ants provide plants with diverse service, such as myrmechory or seed dispersal, besides extrafloral nectaries. In short, keystone resource species provide resources critical to the survival of dependent populations, such as food, shelter, mating site or soil for growth of occupation by social predators of such high caliber. As a result, a large part of the native Hawaiian insects became unfortunate victims of this invasive ant species (Wilson, 2003). Disruption of a well balanced food web in a natural community is, indeed an erosion on the quality of environment.

The endemic taxa, especially in the tropics are particularly susceptible to endangerment and even extinction disrupting the ecological balance since they thrive in restricted habitat patches with specific edaphic, atmospheric and vegetational characteristics. In the event of major global change, they have neither the capacity of shift geographical locations nor the physiological tolerance or resilience to survive at the same location (Samways, 1995). In some respects, the montane and forest endemic insects are less threatened than the beneficial insects of lowland agroecosystems viz., natural predators, parasitoids and pollinators due to indiscriminate introduction for biocontrol and accidental introduction due to quarantine negligence. Besides these aspects, some cosmopolitan species also could go extinct, because they generally rely on habitats that are temporarily available for colonization and act as fugitives that move across the landscape from one habitat to another. Massive alteration of habitats associated with human activities may cause extinction of such species, resulting in a broader geographic effect on comunity formation and ecosystem function, than extinction of local endemics (Cairns, 1993).

CHAPTER 4

INSECT BIODIVERSITY IN XYLOPHAGOUS AND COPROPHAGOUS HABITATS

Changes in habitat patches due to environmental perturbations or human interference are well known and such microhabitats provide favourable niches for a host of insects. With many such habitats becoming increasingly fragmented, they exhibit considerable variation in patch density. Tropical forests are good examples of extreme complexity and diversity with their more dynamic microhabitats whose inhabitants contribute to their decomposition. As such they play a role in the nutrient cycling of the surrounding ecosystem. Most inhabitants therefore have short life times, often due to the action of decomposers resulting in `habitat turnover', whereby old patches disappear and new ones appear. One of the finest examples of such microhabitats relate to dead and decaying logs and dung habitats which provide shelter and resources for their inhabitants. The former is referred to as *xylophagy* and the letter as *coprophagy*. Xylophagous or saproxylous species could be obligatory or facultatively saproxylic being respectively dependent or associated in some part of their life cycle in dead wood. Natural changes occur in dead wood, in the fruiting bodies of fungi and in herbivore dung, which are good examples of heterotrophic microsuccession.

Besides providing a pool of nutrients in the ecosystem they provide a favourable habitat for decomposer organisms. Old, selectively logged forests provide a high value habitat for early successional xylophagous species, despite the lower amount and diversity of dead wood substrate. All classes of decaying wood act as a moisture store providing refugia for ectomycorrhizal fungi. The fruiting bodies of macrofungi such as Agaricales and Polyporaceae provide and excellent environment for several insects. Many invertebrates thrive on decaying wood and other dead parts of plants and perhaps one of the most dynamic ecosystem is dead logs in a forest, acting as ``islands of new life essential for scavengers". Since the remains of plants and animals are storehouses of nutrients, removal of dead logs from natural forest roles the living animals and plants of precious nutrients so essential for life continuity (Kadur & Bawa, 2005).

Fig. 4. Insect damage to felled logs. (After R.V. Varma)

Woody debris provide adequate microhabitat in natural forests for large numbers of insects. Large logs support population of smaller insects for several generations. Dead wood microhabitats or saproxylic taxa, therefore harbour large population of diverse insects, the bulk of them being beetles, flies, ants, wasps, thrips, termites. Being saproxylic they are dependent on some stage of their life cycle on dead and decaying wood or upon wood inhabiting fungi (Fig. 4). Dead wood microhabitats have played a special role in insect evolution, some of the evolutionary novelties in this microhabitat relating to wing polymorphism, besides diversity in freedom and breeding (Hamilton, 1978). Similarly dung beetles are good examples of habitat conservation and ecosystem engineering (Hanski, 2005) with beetles modifying their environment in course of time. Man-made changes in the environment involving habitat loss and fragmentation have become so considerable that their ecological consequences dominate. As such, in recent years there is a shift in research interests and today's concern relates to landscape ecology which deals not only with major habitat types, but also with microhabitats. Apart from factors influencing the risk of extinction of local populations in individual habitat fragments, migration among fragments is an important process for the establishment of new populations. Needless to pinpoint that different species use different kinds of habitat and different species require different amounts of habitat for their presence.

The general trend in most degraded forests is a drastic reduction of species richness. All the same, the timing and periodicity of occurrence of species also vary in relation to the age of the habitat, viz., whether primary, secondary or tertiary. Generally due to a turnover of species assemblages, some decline and disappear, while others tend to increase (Lawton, 2005). In other words, the diversity of species assemblages depend on the diversity of degraded habitats from a localized dead log to surrounding niches, so that patterns become evident as regards the size of degraded habitats. While individual fallen trees show fanistic succession in the course of decay, many forests have been disturbed by selected logging as is evident in the forests of the Western Ghats, not to mention of the sandal wood stocks resulting from tree felling. A logged forest tends to have light gaps providing an environment for butterflies. In many areas where forest fires occur the after effects are of interest in that smaller insects such beetles, mantids, grasshoppers and the like tend to turn black to match the charred trunks. Pyrophilous species are fire favoured occurring exclusively in burnt logs. A good example is *Trypodendron lineatus,* the weevil.

Commerical logging is often seen as a major cause for deforestation, involving some of most valuable trees such as mahogany (*Swietenia macrophylla*) and teak, *Tectona grandis*. Such selective logging brings down many nearby trees causing damage to the forest, reducing biological diversity considerably. Such logging results in scattered fragments or islands, poor in species or biological richness. As rainforests become increasingly fragmented, such interdependence will crumble for the lack of keystone species such as ants. An important microhabitat relates to the basal parts or stumps of logged trees which, overtime habours a lot of insects, notably ant colonies and beetles. Rotting logs and stumps in all stages of decay therefore provide habitats for several insects. Dead wood and other habitats types for saproxylic fauna should be retained, since the diversity of habitats even within an old forest plantation becomes important to maintain the diversity of arthropod assemblages.

Downed logs are important as colonizing substrate for ants, acting as agents of wood decay. Stages of decay are associated with arthropod invaders inoculating the tunnels with fungi, bacteria and phoretic mites and subsequent feeding by beetles enabling tunneling in the logs, opening further access to other insects and arthropods (Stevens, 1997). As such the maintenance of natural diversity of species through dead and decaying wood and debris not only enables habitat renewal, but also sustains biodiversity (Fig. 4.1 & 4.2).

Dead wood, woody debris microhabitats

A fallen tree is a home for a host of insects. With the tree beginning to show a slow process of decay at the early stage it tends to be attractive to a wide range of wood feeding insects such as the long horn beetles, the females biting holes in the bark to lay eggs, followed by the weevils. As the beetle larvae get to the bark on the fallen timber, a new set of opportunities arise with beetles such as bostrychids (species of *Dinoderus, Sinoxylon, Lyctus, Heterobostrychus,* platypodids *(Crossotarsus, Platypus), scolytids* (several species of *Xyleborus*) and termites such as *Microcerotermes, Odontotermes* succeeding one another. The decomposition of wood occurs through the combined activities of fungi and insects. Collembola, Diptera, Coleoptera most frequently occur in woody fungi, while Diptera are predominant in fleshy agarics and polyporous fungi. Some polyporoid or bracket fungi are known to harbour more than 250 species of insects like *Fomes fomentarius,* a polyporous wood inhabiting fungus causing rot of heart woods.

Fig. 4.3. Wood infesting fungi (Top) *Schizophyllum* (Bottom) *Psathyrella* (After K. Natarajan)

The sporophore community comprises major mycophagous species, predators, scavengers and tunneling insects like nitidulid and staphylinid beetles. Fleshy fungi are inhabitated by Diptera such as Phoridae and Anthophoridae, a few Drosophilidae and Tipulidae. Siricid wood wasps *Sirex, Tirocerus, Tremax* which feed on the fungus *Amilostereum* are well known, possessing a long ovipositor which penetrates the bark and into wood. Some species of *Sirex* have hypopleural organs on the first abdominal segment of the female, habourinmg the fungus *Stereum sanguinolentum*. Similarly mycetophilids in relation to dead wood habitats are well known, having coevolved with various groups of fungi. Several mycetophilids are adapted to living in dark humid habitats. The diversity of Mycetophilidae in a site is directly related or influenced by fungal diversity, since most of the species have their larval development in fungal habitats (Okland, 1994). The number of mycetophilid species is clearly low in cut forests, since several fungal species disappear. Several dead wood variables such as moss covered dead wood, decayed dead wood etc have shown the presence of mycetophilid flies.

Fig. 4.4. A fallen tree infested with fungi (left) *Marasimus* sp. (right)

Several fungi abound in the fungus garden of termites the most abundant being *Termitomyces* of which there are several species which appear soon after rains. The termites emerge to the surface to collect fungi (Fig. 4.5). Closely resembling fungus-growing termites are the fungus growing ants, which are believed to represent the most advanced stage in the evolution of fungus gardening because they feed only on the fungus and cultivate it (Batra and Batra, 1967). The fungus gardens also consist of clusters of bromatia, the swollen tips of the filaments forming the mycelium. Bits of bromatia are fed to the larvae that grow in the mycelium of the fungus garden. The excreta

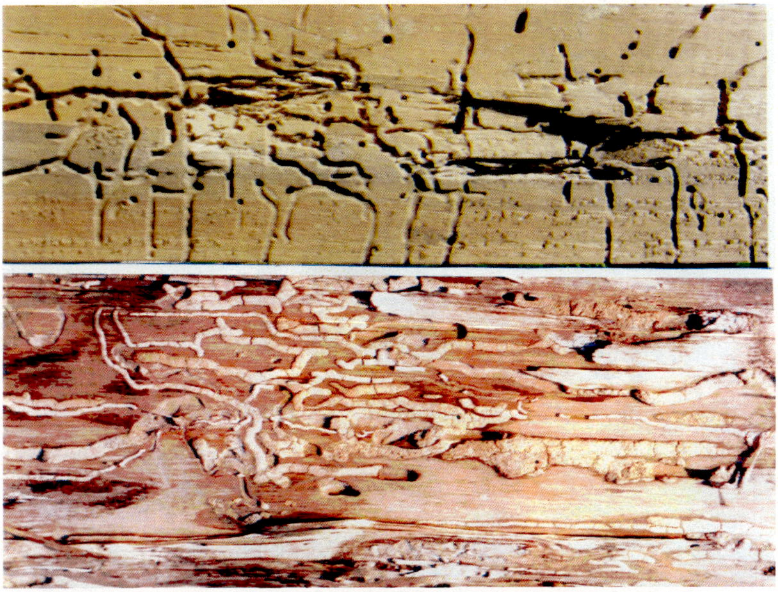

Fig. 4.1. Top : Powderpost beetle damage (Bottom)

D. ocellaris

D. minutus

Sinoxylon anale

Heterobostrychus aequalis

Fig. 4.2. Powderpost beetles (After R. Sunderaraj)

Fig. 4.5. *Termitomyces* (After K. Natarajan)

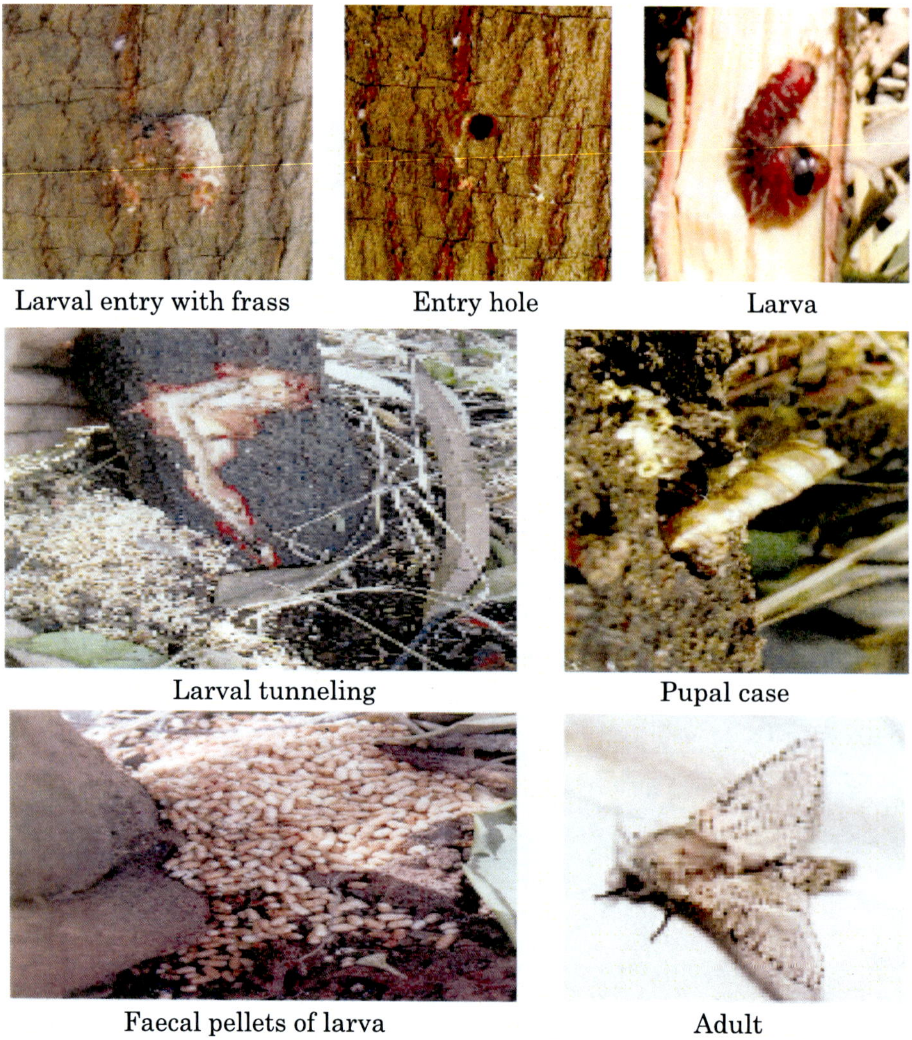

Fig. 4.8. Infestation of red stem borer on sandal (After R. Sundararaj)

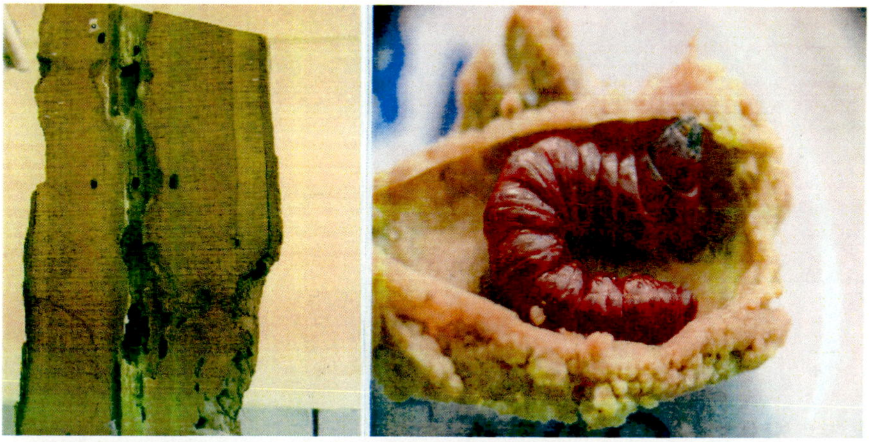

Fig. 4.9. Damage and Adult emergence holes of *Batocera rubus* (after R. Sunderaraj)

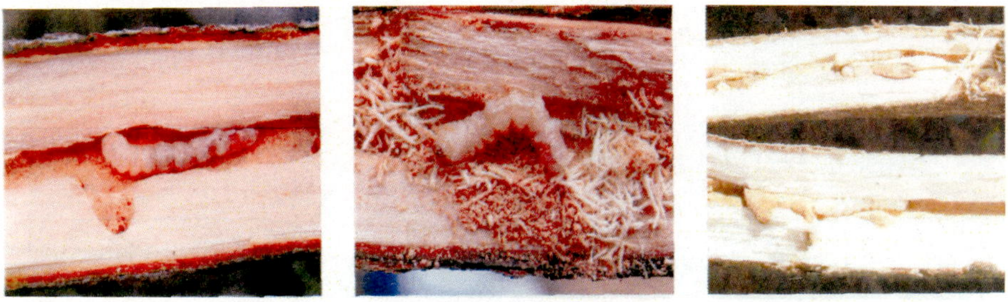

Damage of sandal heartwood by the grub

Galleries made by grub | Exit hole-inside view | Adult emergence

Exit hole-outside view | Adult on branch | Adult feeding marks

Fig. 4.10. Infestation of Sandal heartwood borer, *Aristobia octofasciculata* Aurivillius

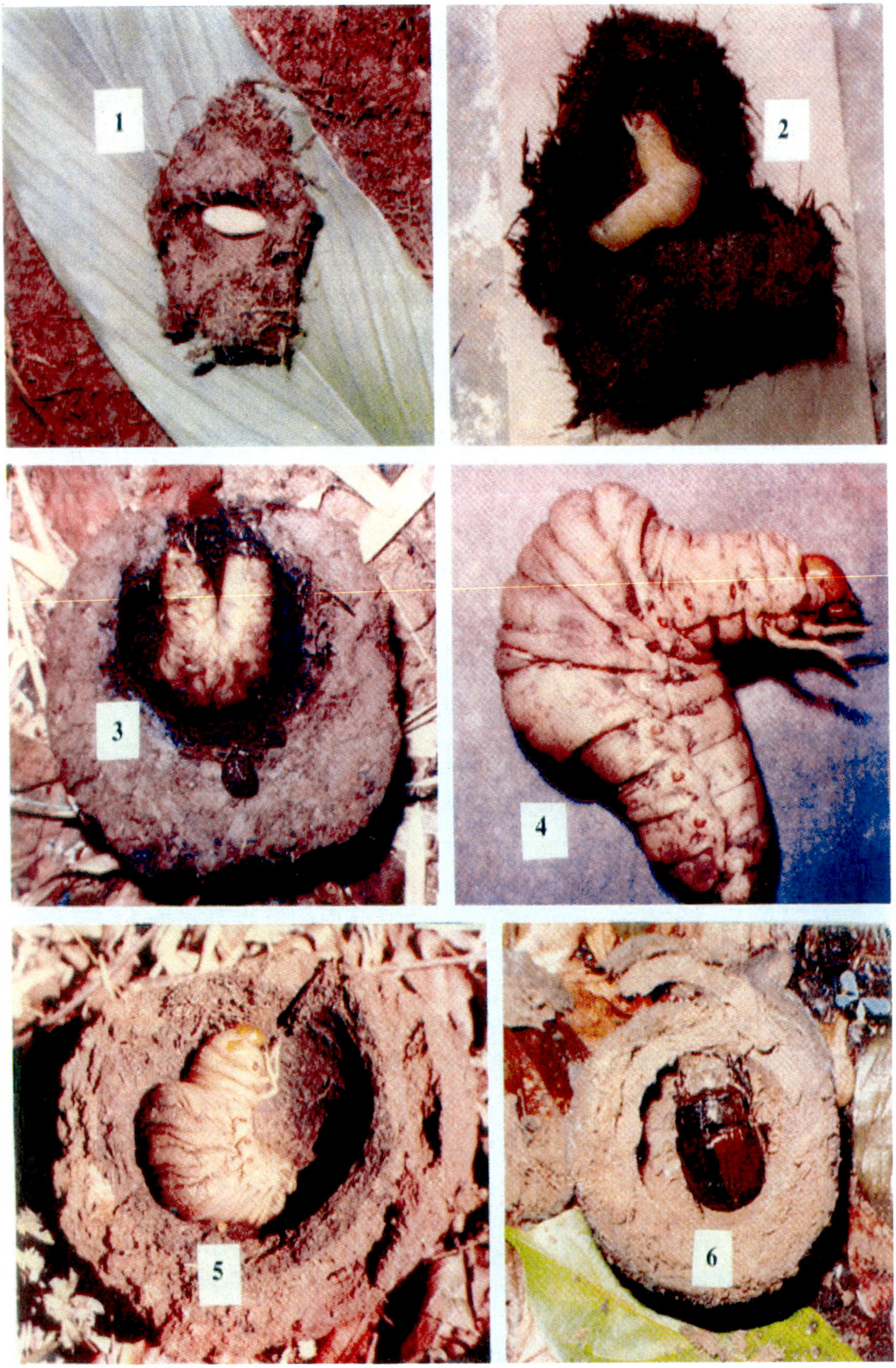

Fig. 4.18. Developing stages of *Heliocopris dominis*
(After K.J. Joseph)

and saliva of ants deposited in the fungus garden influences the growth of fungus. Some species of the genera *Speculitermes, Pericapritermes, Anoplotermes* also feed on soil and humus bringing about modification of the soil (Sen-Sarma, 1996). *Reticulotermes, Coptotermes, Macrotermes* establish large colonies in the soil or above ground and foraging workers attack wood often in direct contact with the ground. *Heterotermes indicola* in forest and urban environments with underground nests form large colonies and are the most destructive of termites in India, causing considerable loss to wood. As such the contribution of termites to the decomposition of wood biomass in tropical forests is very significant, their activities tending to cause modification of soil profiles and fertility, through bringing large quantities of subsoil to the surface. The functional group of termites are soil feeders, soil/wood interface feeders, woodfeeders and litter foragers (Watt, Stork and Hunter, 1997).

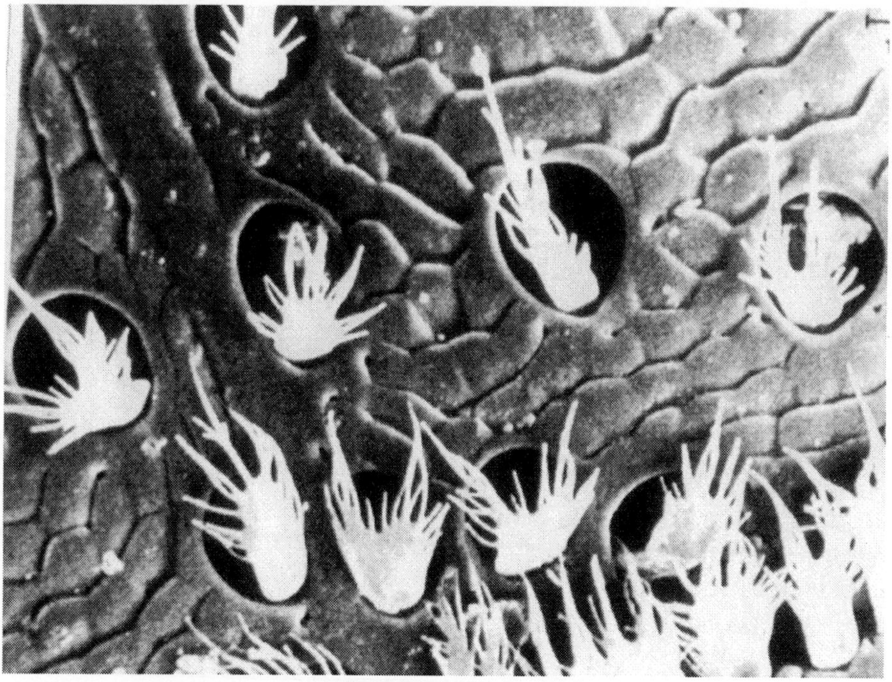

Fig. 4.6. Fungal hyphae germinating from mycangia

Many trees belonging to the Santalaceae, Myrtaceae, Fabaceae, Meliaceae and Combretaceae are susceptible to termite attack and damage, and ponerine ants are efficient predators of termites. Such species of ponerine ants as *Leptogenys processionalis, Diacamma rugosum, Harpegnathus saltator, Tachycondyla, Tesserinoda* are among the common species predatory on termites, specially the species of *Odontotermes* (Shivashankar *et al.,* 1995).

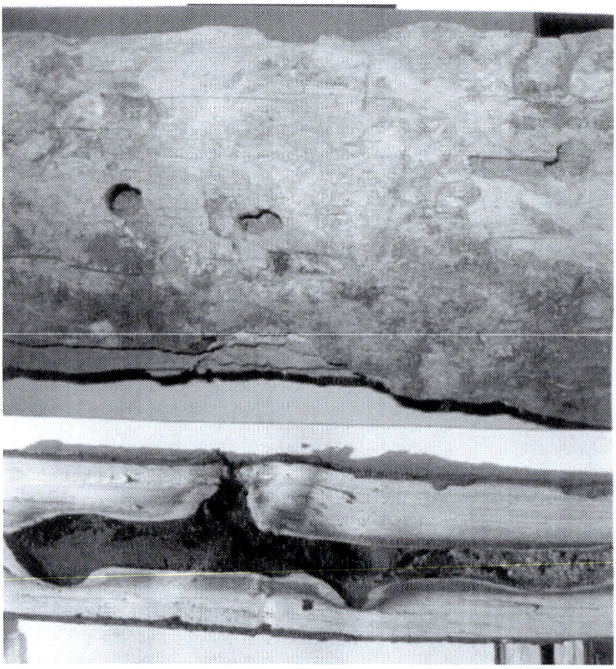

Fig. 4.7. Tunnel made by *Indarbela quardinotata* in Sandal

The decomposition of wood occurs mainly through the combined activities of insects and fungi. Mathur and Singh (1960) provide a long list of insects infesting teak, *Tectona grandis* including over a hundred species, mostly coleopteran boring into dead sapwood. High infestation of cerambycid beetles, with tunnels reaching as far as the heartwood is very typical of the logs of *Palaquium ellipticum, Terminalia alata, Diospyros* sp, *Callicarpa tomentosa* and *Ponophiolus javanus.* Staphylinids occur in highly degraded wood, besides passalid beetles *Leptoulax dentatus* occurring within bores in rotting or degraded wood. Cerambycid beetle tunnels are also infested with the hysterid beetle *Platysoma atratum.* Platypodids or ambrosia beetles bore deeply into wood of felled trees, logs or stumps, cultivate such ambrosia fungi like *Lasiodiplodia, Monilia, Cephalosporium* in the galleries and feed on them. The ambrosia beetles and the fungi in the fallen logs have had special importance from the viewpoint of the ecology of these beetles and their nutritional aspect, with a view to protect the wood material of logs and felled trees. The ambrosia beetles have within their bodies glands called *mycetangia* in which they store and protect the spores. The majority of species belong to the Scolytidae and Platypodidae, the former including several species of *Xyleborus, Xylosandrus, Euwallacea* and the latter, species of *Platypus, Crossotarsus* to mention the more common ones. Scolytids or bark beetles are secondary exploiters, often attacking weakened trees and have great impact on forest communities in terms of stocking and harvesting timber. The ambrosia fungi are highly species specific in relation to the insects and flourish inside the wood tunnels throughout the active life of the beetles. When the ambrosia beetles tunnel into wood, spores are dislodged from the Mycangia. The interior of the tunnel soon develops a mass of velvety fungi and on this beetles feed (Fig. 4.10).

Equally important from the economic point of view are beetles boring into piled stocks of sandalwood and bamboos, the latter in and around paper factories. Wood-infesting beetles have attained certain anatomical and physical adaptations to feed on dry wood. In general, the larvae of Cerambycidae are large, long and elliptical, the larvae of Anobiidae and Lyctidae are small curved and grub-like; the larvae of Buprestidae are large and tubular, with a flat, oval thorax. Species of *Xylocopa* carpenter bees viz *X. aestuam, X. iridipennis* nest on dead logs, while the lucanid *L. cervus* develop in decaying wood. Pselaphids occur in rotten wood eg. *Mastiger abruptus*. Also found as wood feeders are the cockroaches *Panestia* and *Cryptocerus*. Stocked mango wood is affected by the curculionid *Camptorrhinus mangiferae* and plywood by *Heterobostrychus acquilis*. Passalid beetles occur in very large numbers, often gregariously, on the floor of wet tropical forests making tunnels inside logs of fallen trees, wherein larvae live and feed resulting in the disintegration of wood. The teak borer *Alcterogystia (=Cossis) cadambae* causes external bark injury riddling the bark with numerous holes and fungal infestation results in wood decay (Fig. 4.10). One of the largest of cerambycids *Hoplocerambyx spinicornis* attacks felled trees, their grubs forming galleries and large scale epiderics are known. The larvae of cerambycids being borers beneath the bark of recently felled hardwood trees, attack wood only when the bark is present. Their loud chewing sounds have earned them the name ``sawyers''. Young larvae feed beneath the bark forming shallow galleries between the bark and sapwood. Older larvae bore into hardwood and bore to the surface, so that the feeding galley extends from the bark to the heartwood and bark again (Robinson, 1996). While the cerambycids are known to attack fresh dead wood with the intact bark, anthribid, curculionid and lyctid beetles infest wood beneath bark. When the ambrosia beetles tunnel into wood, spores are dislodged from the mycangia. The interior of the tunnel soon develops a mass of velvety fungi and on this the beetles feed (Fig. 4.11).

The cavernicolous quality of insects living in spaces which dead trees offer, afford excellent breeding sites for diverse insect specis, providing a good example of functional convergence (Hamilton, 1978).

The habitats lying immediately beneath bark are extremely productive examples of wing polymorphism, of male haploidy, special breeding structures and development of certain types of sex-limited diversity. The condition of the bark, the degree of decay as well as the state of invasion of fungi are essential factors, with very often a series of fungi attacking in succession enabling a confirmed availability of mycelia or spores. Many species such as those of mycophagous thrips which readily tend to mate within the habitat improving the potential for increase and colonization, contributing to the development of subsocial behaviour (Ananthakrishnan, 1984). Besides the bark infesting psocids like species of *Liposcelis* several species of mycophagous tubuliferan thrips abound in the bark, feeding on both mycelia and fungal spores. Mycophagous species such as *Hoplothrips fungosus* utilize the fruiting bodies of *Polyporus* species, the females laying eggs within the comb-like structure of the fungus. Other species of *Hoplothrips* are known to inhabit shelf fungi such as *Polyporus, Ganoderma and Stereum.* Besides several species of tubuliferan thrips occur in the bark of dead trees depending on fungal hosts. *Bactrothrips idolomorphus* feeding on the spores of *Pestalotia, Dinothrips sumatrensis* on *Lasiodiplodia theobromae, Elaphrothrips procer* on *Pestalotia* and *Lasiodiplodia* on *Tectona grandis* and *Shorea robusta,* to mention a few. Mycophagous thrips inhabited niches also reveal the presence of pathogenic fungi as *Verticillum, Fusarium, Lasiodiplodia, Pestalotia, Phomopsis* etc.

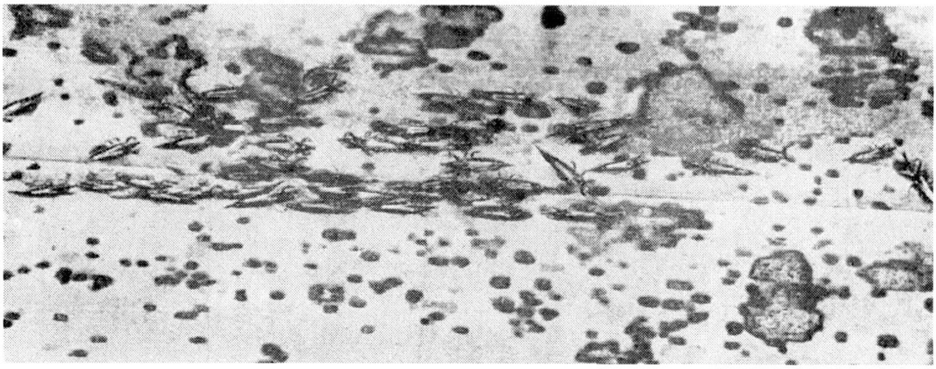

Fig. 4.11. Colony of *Tiarothrips subramanii* on drying palmyra leaf

In view of the undisturbed nature of the microhabitats the mycophagus species feeding on saprophytic fungi abound in forests, in particular the tropical rain forest or evergreen forest belts of this country comfined generally to South-Western parts extending all over Kerala to Goa (Western Ghats), parts of the Eastern Ghats and the Subhimalayan forest ragnes. As such, the nature of the forests and related changes in rainfall, temperature and vegetation naturally result in variation in the periodicity and abundance of insects like thrips which occur in specific nickes like dry, fungus infested leaves either on fallen trees or branches and underneath bark.

Fig. 4.12. Mycophagous *Bactrothrips* (B) and *Dinothrips* (A,C)

Fig. 4.13. Fungal diversity in relation to mycophagy in thrips.

Dense aggregation of *Ecacanthothrips tibialis* occur on fallen twigs and drying bark sometimes in densities of 180-200, individuals. Drying sheaths of *Areca catechu* show *Stigmothrips, Azaleothrips, Ecacanthothrips* which are mycophagous, besides *Elaphrothrips* and *Mecynothrips. Mallotus, Anacardium, Hevea* show aggregates of *Dinothrips sumatrensis* under bark of stocked wood, while *Meiothrips menoni,* a less aggregate species (17-24) occur in leaf litter of *Ocna wallichi. Bactrothrips idolomorphus* occur in dry litter of *Shorea robusta* forming dry aggregates. *Tiarothrips subramanii* occurs in dense aggregates in drying leaves of *Borassus flabellifer*. A diversity of fungal habitats may be required to support the full diversity of thrips and saproxylic beetles. Interestingly enough (Mound, 2005) when rearing *Hoplothrips fungosus* on *Stereum* fungus on dead wood, it was found that the thrips colony remained healthy with all adults flightless and when this fungus was overtaken by *Mucor*, the colony produced a large number of winged adults of both sexes.

Fig. 4.14. *Ecacanthothrips tibialis* (extreme males) and larvae feeding on spores

In dense, low forests with trees at heights of 16-20 meters with much under growth and many epiphytes, trees with mosses and ferns, also harbour some species of thrips. In open stunted thorn forests with rainfall under 65 cm, the trees are shorter and more scattered, with acacias more predominant, as also the palm *Phoenix sylvestris,* besides xerophytic bushes which harbour some common species. As such the varied nature of the forests and the related changes in rainfall, temperature and vegetation result in variation in the periodicity and abundance of thrips which occur in specific niches like the drying fungus infesting leaves on fallen trees and branches, in the leaf litter forming the forest floor, underneath bark etc. Several species are restricted to feeding on fungal hypae (mycetophagous), while many others feed on spores (sporophagous). Thrips inhabited niches reveal the presence of potential pathogenic fungi such as *Verticillium, Fusarium, Trichothecium, Lasiodiplodia, Pestalotia,* etc. Some of the common species of mycophagous and sporophagous thrips are :—

Mycetophagous	Sporophagous
Azaleothrips amabilis	*Bactrothrips idolomorphus*
Ecacanthothrips tibialis	*Dinothrips sumatrensis*
Hoplandrothrips flavipes	*Elaphrothrips procer*
Neurothrips indicus	*Meiothrips menoni*
Polyphemothrips craceus	*Mecynothrips simplex*
Stictothrips fimbriata	*Nesothrips acuticornis*
Stigmothrips limpidus	*Gastrothrips falcatus*
Several species of *Xylaplothrips*	*Tiarothrips subramanii*

Dead logs in forests therefore form one of the most dynamic ecosystems `acting as islands of new life essential for scavengers that take what they need while returning other nutrients to the soil. Decomposition of dead material releases populations of decomposers scavengers and other organisms surviving on them.

Ecological implications of xylophagy

Ecological processes in forests relate to building as well as break down processes. Large woody material contains sufficient carbon and energy contributing to the food Web. Following the building phase, there is the decaying or breakdown phase in wood in large or small pieces of dead, coarse woody debris. Rotting or decaying wood forms an important component in the build-up of soil structure, nutrient cycling and habitat for soil organisms. Some of these components include rotten logs or stumps of wood, source materials for soil build-up. They play a useful role in the productivity of forest trees as well as providing habitat and structure enabling maintenance of biodiversity. All size classes of decaying wood pieces contribute to the carbon pool and act as a moisture store providing refugia for ectomycorrhizal fungi, besides another component, the lichens enabling provision of a good habitat.

During the early stages of decomposition mushrooms growing on decaying logs serve to increase concentration N K P (Sodium, Nitrogen and Phosphorous) in the logs adding to the nutrient pool in the debris. Anthropod and earthworms digest the nutrients and return them to the forest through their frass. Thus coarse debris when added to the ecosystem at regular intervals provides a long term source of nutrients (Stevens, 1997). Therefore, removing large pieces of decaying wood tend to alter the components of a forest. As such coarse woody debris plays an important role in creating habitats for many species of plants and animals. They form habitats for microbial decomposers bacteria fungi, an energy source for a complex food web, protective cover for inhabitants providing moist microsites for the fauna.

In short the woody debris plays a useful role in providing (a) favourable habitats for microbial decomposers such as bacteria and fungi, forming an energy source for a complex food web (b) hiding cover for insects and their predators (c) a food base microsite for fungi (d) a useful colonizing substrate for ants, besides other soil micro-arthropods, useful in nutritional cycling.

The stages of decay create varied habitats over time for several organisms including insects. To quote Stevens (1993) ``Arthropod invaders inoculate their tunnels with fungi, bacteria, phoretic mites, nematodes and Protozoans; later the beetles that feed on sapwood and heartwood tunnel deeper into the log, opening access for other arthropods. These processes involve hundreds or possibly thousands of species and are a critical link in the carbon and nutrient cycles of the forest."

Pests of stored bamboos

Family : Bostrychidae

1. *Dinoderus minutus* Fab
2. *D. ocellaris* Steph
3. *D. brevis* Horn

4. *D. japonicus* Lense
5. *D. pilifrons* Lense
6. *Bostrychopsis parallela* Lense
7. *Sinoxylon anale* Lense
8. *Heterobostrychus aequalis* Waterhouse
9. *H. unicornis* Waterhouse

Family : Lyctidae

1. *Lyctus africanus* Lense
2. *L. brunneus* Stephens
3. *L. sinensis* Lense
4. *Minthea rugicollis* Walker

Family : Curculionidae

1. *Conarthrus affinis* (Attack half dead and dead bamboos)

List of timber beetles collected from dead or fallen wood on forest floor in Kerala (after R.V. Varma)

Insect family/species	Host timber	Remarks
Family : Cerambycidae		
Chloridolum alcmene	*Palaquium ellipticum*	High infestation with tunnels reaching
Coptops sp.	*Diospyros* sp., *Terminalia alata*	Gallery reaches as far as the center of the log. Minor damage.
Glenea sp	*Palaquium ellipticum*	"
Myagrus hynesia	*Ficus glomerata*	Low infestation Damage minor.
Xylotrechus subscutellatus	*Callicarpa tomentosa*	Bores into the heartwood.
Family : Bostrychidae		
Sinoxylon atratum	*Acacia* sp	Infestation high. Cause serious damage to sapwood.
Scolytidae *Ambrosiodmus* sp.	*Acacia* sp.	Infestation high. Cause serious damage to sapwood.
Diamerus curvipes Wlk.	*Persea macrantha*	Low infestation confined to the outer sapwood.

Euwallacea andamanensis	*Myristica malabarica*	"
Microperus alpha	*Cinnamomum* sp., *Terminalia* sp.	"
Sphaerotrypes coimbatorensis	*Myristica dactyloides, Agrostistachys meeboldii*	High infestation restricted to bark.
Xyleborus andrewesi	*Ficus* sp.	Infests live trees. Minor infestation.
Xyleborus perforans	*Grevillea robusta, Terminalia bellerica*	Infests live trees. Gallery reaches as far as the center of the log.
Xyleborus assamensis	*Myristica malabarica, Schleichera oleosa Tectona grendis*	High infestation causing extensive damage to the bark and outer sapwood. Larvae in dead sapwood.
Xyleborus butamali	*Cinnamomum* sp.	Minor attack confined to the outer sapwood.
Xyleborus cognatus	*Terminalia* sp.	"
Xyleborus morstatti	*Palaquium ellipticum*	"
Xylosandrus crassiusculus	*Myristica malabarica*	"
Xylosandrus terminatus	*Grevillea robusta*	Gallery reaches as far as the center of the log.
Xylosandrus sp.,	*Bischofia javanica*	Low infestation Attack confined to the sapwood.
Platypodidae		
Crossotarsus saundersi	*Myristica malabarica*	"
Crossotarsus lunatus		
Crossotarsus minax	*Palaquium ellipticum*	Gallery reaches as far as the center of the log.
Crossotarsus squamulatus	*Schelichera oleosa*	Gallery reaches as far as the center of the log.
Diapus impressus	*Palaquium ellipticum, Terminalia bellerica, Mallotus philippensis*	Gallery reaches as far as the center of the log.
Diapus quinquespinatus	*Terminalia* sp.	Minor attack confined to the outer sapwood.
Platypus biformis	*Bischofia javanica*	Minor attack confined to the outer sapwood.

Platypus cupulatus Chap.	*Bischofia javanica*	Minor attack confined to the outer sapwood.
Platypus cupulifer Wich.	*Ficus* sp.	Minor attack confined to the outer sapwood.
Platypus cylindrus Fb.	*Palaquium ellipticum*	Gallery reaches as far as the centre log.
Platypus sp.	*Terminalia bellerica*	Minor attack confined to the outer sapwood.
Family : Curculionidae		
Hylobius angustatus Fst.	*Unidentified timber*	Minor damage to outer sapwood.
Camptorrhinus mangiferae	*Dalbergia latifolia*	Bores deep into the wood, with galleries reaching as far as the herartwood.
Mecistocerus simplex	*Callicarpa* sp	Mild damage.
Crytorrhynchus rufescens	*Ficus* sp., *Canarium strictum*	"
Trachodes sp. *nr. acutangulus*	*Myristica malabarica*	Mild damage.
Family : Passalidae		
Leptaulax dentatus	*Unidentified decaying wood*	Borers in rotting wood, making large cavaties.
Family : Histeridae		
Platysoma atratum	*Terminalia* sp., *Ficus* sp.	Beetle collected from cermbycid borer tunnels; probably predating on larvae.
Family : Staphylinidae **Subfamily : Paederinae**		
Pinophiolus javanus	*Myristica malabarica*	Found within highly degraded wood.
Order Dictyoptera Family : Blattidae		
Dicellonotus sp.	*Palaquium ellipticum*	Bores deep in the rotting wood making large excavations.

Fig. 4.15. Some coprophilous beetles (After Sabu Thomas)
A. *Gymnopleurus melanarius* with a dung ball, B. Elephant dung balls (Courtesy: R. Sukumar, 2003), C. & D. Dung beetles associated with elephant dung. C. *Heliocopris dominus* (Exclusive elephant dung tunneler), D. *Onthophagus vividus*

DUNG BEETLE MICROHABITAT

Dung beetles exploit ephemeral dung parts as resources and participate in the recycling of matter, nutrients and energy, contributing the natural process of regeneration of ecosystem. Species of *Sisyphus, Symnopleurus, Liatongus, Oniticellus, Copris and Heliocopris* occupy diverse ecological niches due to their foraging tactics, depending mostly on dung produced by herbivorous mammals like the Indian bison or gaur, the elephant, grazing cattle and the like. Such resources being very patchy in space and time, resource partitioning and competition become essential features of dung beetle assemblages. Strong competition communities. Ecology, community structure, density and abundance, successional trends and resource partitioning of dung biotic communities become important.

Fig. 4.16. Scarabaeids feeding on dung, from Western Ghats (After Sabu Thomas), a. *Sisyphus neglectes* b. *Sisyphus longipus* c. *Drepanocerus setosus* d. *Liatongus indicus* e. *Oniticellus cinctus* f. *Copris davisoni* g. *Copris repertus,* h. *Onitis siva* , i. *Onthophagus bronzeus*

Two types of dung beetle communities are observable on the forest floor-those associated with larger parts of elephant and gaur and those with smaller droppings of deer and boar. Further, dung beetles are classified broadly into their functional groups based on their nesting strategies *rollers* (helecoprid nesters), *tunnelers (*paracoprid nesters) and *dwellers* (endocoprid nesters). Rollers form balls from the dung pat, which are rolled away, buried and used for feeding and breeding. Tunnelers create underground chambers beneath the dung pat and construct their nests using dung from the pits, whereas dwellers breed in the dung pat itself. This functional stratification enables dung beetles to minimize the intense competition for limited food space and also protect the food from adverse environmental conditions. Based on the order of arrival and rate of resource utilization of dung pats, dung beetles are grouped into Type 1 or Type 2 categories, the former being more superior competitors adapted to use fresh dung as fast as possible, while the latter are adapted to use older dung (Hanski, 1991, Sabu Thomas *et al.,* 2006). Age, size, water content and dung structure determine the preference of dung beetles towards a specific dung type. The surface crust prevents moisture loss from the bottom layers in ruminant dung, enabling prolonged stay of some species. In the elephant dung no such crust formation occurs and the coarse texture of the dung is preferred by some than the more fluid, fine trextured dung of ruminants. The presence of *Aphodius* is may in directly influence the degradation of cowpats by altering their attraction to subsequent colonizers (Holton, 1979).

Biology and Breeding behaviour

Heliocopris dominus depends exclusively on the fresh dung of elephants for feeding and breeding. The development of this beetle from egg to adult takes place inside a brood ball, lodged in a brood chamber located in the soil at a depth varying from 27-56 cms. New generation adults commences emerging from the soil after heavy rains, when the brood balls in which they have developed, as well as the soil above are adequately water-soaked (Joseph, 1998). Beetle emergence takes place after nightfall, normally from 6-9 p.m. Emerging beetles soon fly to fresh elephant dung pats to which they are strongly attracted by the dung odour. After a period of maturation feeding lasting approximately for 4 weeks inside separate feeding tunnels, adults of sexes emerge (Figs. 4.17; 4.18).

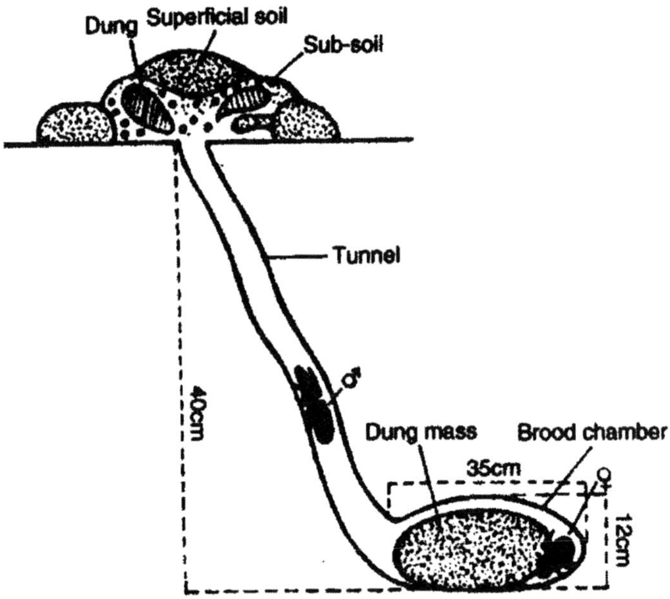

Fig. 4.17. Vertical section through dung pat (After K.J. Joseph)

In nests more than 15 days old, the females start brood ball construction. The compact dung mass is cut up with submasses, each of which is eventually incorporated with a brood ball containing a single egg, the number of brood balls in a brood chamber varying from 2-4. Some of the brood balls are spherical, while a few others are subspherical, somewhat pear-shaped. As a result of highly organized breeding behaviour and nidification including well provisioned brood ball construction, the brood develop in an environment mostly isolated from fluctuation of microclimate and natural enemies. Further, the mother continues to live inside her brood chamber, taking care of the developing larvae, until they pupate. This active maternal care become significant in terms of brood survival, leading to reproductive success (Joseph, 1998). The egg of *Helicopris dominus* appear to the largest of insect eggs. The larvae go through three instars, the first and second lasting two months each. The prolonged developed of the third instar takes place in two stages, an active feeding stage lastly 3-1/2 months in the west season and an activity, quiescent stage in the dry season. The adult emerge during the wet season and there is only one generation annually (Joseph, 2003). The adults remain inside the brood balls until sufficient rain water penetrates into the brood chamber. Subsequently each beetle cuts a hole and escapes. There has been a gradual elimination of the dung microhabitats, thanks to human impact on the taxa with improved pastures and modernization of animal husbandry. As been stressed by Hanski (2005) ``the loss of the dung microhabitat in agroecosystems illustrates very clearly the consequences of microhabitat loss and the way in which biodiversity and ecosystem functioning are related". Besides parks, urban forest and garden are typically cleared of dead and drying trees thereby eliminating the microhabitats close to the soil surface. Deforestation, forest fragmentation, logging and farming tend to intensify climate change. Even then such degraded habitats lose species and the loss is not uniform across all taxonomic groups.

Insects tunnel through dung increasing the area available for microorganisms and enhance degradation through fragmentation, aeration, burying or mixing large quantities of dung with surrounding soil. For example the presence of *Aphodicus* may indirectly influence the dehydration of cow pats by altering their attractiveness to subsequent colonizers (Holter, 1979). Arboreal dung beetles are also known foraging in the forest canopy, a guild of beetles inhabiting metres above the forest floor, feeding mostly on private dung. Several species of *Onthophagus* besides others make and roll balls from monkey dung.

CHAPTER -5

CANOPY ECOLOGY AND INSECT BIODIVERSITY

The uniqueness of tropical forest canopies has come to be increasingly realized in recent years and have often been referred to as the `powerhouses of forests', where most of the photosynthesis occurs and where 95% of the biomass is produced. Canopy science has emerged as an indisciplinary, integrated subject relating to the structural and functional aspects involving community architecture as well species composition, nutrient cycling and animal plant interaction. The forest canopy has been defined by Nadkarni (1994), a renowned canopy scientist as the ``aggregate of all crowns in a stand of vegetation, which is a combination of all foliage, twigs, fine branches and epiphytes in a forest". In short, the components of a canopy involve the tree crown, epiphytes, insects and other organisms. Besides, fallen leaves wherein insects may change soil fertility and insect feeding on foliage of trees can influence the quality of forest soils by the dropping faeces or frass (Rinker *et al.,* 2001). Arboreal microhabitats are three dimensional, enabling studies on vertical stratification among arboreal taxa, with arboreal ants exhibiting vertical stratification, many being restricted to high canopy (Longino and Nadkarni, 1993). Interestingly enough one of the early discoveries relate to thrips as pollinators of towering dipterocarp forests (Chan and Appana, 1980).

Canopy research has not taken off in India, though isolated attempts are being made to study canopy trees pollinated by social bees, followed by community level exploration of pollination modes in the wet forests of Kalakkad Mundanthurai Tiger Reserve (Devy and Ganesh, 2003). The scope for canopy research in India with its lush forests such as those of Western Ghats and Himalayan sub regions, appear considerable, and need early attention through interinstitutional involvement with adequate support from funding organizations. Even as early as 1878, A.R. Wallace, has indicated that ``overhead, at a height of hundred feet, is almost an unbroken canopy of foliage formed by the meeting together of great trees and their interfacing branches and their canopy is usually so dense. It is a world in which man seems and intruder and where he feels overwhelmed." Margaret Lowman and Nalini Nadkarni are famous names in the field whose outstanding contributions in this field have laisd a solid foundation for canopy research. Besides the most complex interaction the threatened extinction of canopy species has provided incentives to canopy research. To quote Lowman and Maffat (1993) ``Canopy ecologists are now designing sampling protocols to account for spatial and substrate heterogeneity of their environment. The diversity and distribution of major substrates in the canopy ecosystem are not adequately documented. With the growth of the canopy overtime, canopy branches grow with increased complexity of communities within them". Canopy research has therefore emerged as a new dimension to our study of ecosystems, though with habitat fragmentation and deforestation, one has to cope up with the prospects of adequate loss of information on canopy arthropods (Fig. 5).

Fig. 5. Canopy and ladder climbing

Since no specific attempts have been made to study canopy arthropods in India notably with respect to insects, a taxonomic appraisal of canopy fauna becomes impossible. However the degree of diversity of canopies of forests of diverse types has been so efficiently documented by Stork *et al.* (1997) and which serves as a pathfinder for future work in this direction. A broad outline of canopy research explaining diverse aspects of canopy ecology and faunistics are available in the works of Stork, Nadkarni and Lowman, wherein some basic trends are presented as a guidance for future research in this area. Canopy fogging with knock down effect with insecticides has revealed unexpected richness of canopy arthropods, providing the basis for fairly relevant estimate of the number of insects. Most arthropods are extremely sensitive to very low concent-rations of pyrethrum varies with regard to the type of forest and short exposure times makes this technique highly sensitive for study of canopy arthropods.

Fig. 5.1. Canopy climbing (After R. Sunderaraj)

The number of individual insects collected on various trees hosts varies with regards to the types of forests and available data indicate more a hundred thousand individuals belonging to diverse insect orders collected from varying heights. Studies of beetle diversity have been more focused in relation to their effective specialization in tropical forest tree species (Stork *et al.,* 1997). Insect species richness in an area tends to be largely dependent on plant species richness, as well as the number of species per host plant species and the level of host specificity. For example, different species of Dipterocarpaceae (sp. of *Shorea*) have variable population of beetles, not to mention of faunal variability with in trees with dark dense canopy. Erwin (1983) has indicated that in four different habitat types in Brazil, he found 83% of the 1080 beetles species being restricted to only one forest type. As such, the need for an assessment of trophic guilds such as fungivorus, saprovores, herbivores, predators and xylophages become relevant. Canopy species tend to get reduced or fragmented thereby eliminating species due to habitat destruction. Many have already disappeared, so that an understanding of species diversity in tropical habitats becomes a necessity, before the remaining species are eliminated.

Erwin (1982, 83) has estimated about 600 species of insects specialized to each tree species or genus groups of tropical trees. Since data on arthropods of various parts of tropical trees in India are lacking, examples from other tropical forests may not seem irrelevant. Estimates from Indonesian forests have shown there are 4200 individuals/sq. metre distributed among the canopy, tree trunk and litter, with 1200 insects in the canopy, 50 in the tree trunks, ten in ground vegetation and under 70% of all individuals they found in the soil and litter, mostly Collembola and mites. The consolidated information on canopy arthropods by Stork, Adis and Didham (1997) relating to the Amazonian, Ruandan, Sulawesi, Borneo, Indonesia, New Guinea and Malaysia is undoubtedly an encyclopaedic literature which tends to be an eye-opener for such work in the Indian context.

The number of dominant species supported by each tree species on various trees varies and is correlated with the nature of food resources available, not to mention of the fact that certain proportions of insect species are able to exploit different habitats, the population abundance of generalists tending to be greater than specialists. Maximum biomss attained by any insect species on each tree species appears to be influenced by the leaf expansion rate of young leaves, with insects attaining high biomass on young foliage which grows over prolonged periods (Basset, 1997).

Canopy life generally implies behaviour and morphological preadaptions, which overtime gives rise to forms with increased adaptability in terms of mobility and ability to stay in the canopy, which is therefore a structured ecosystem with interdependent species. The diverse microhabitats and types of substrate existing in the canopy, contribute to canopy biodiversity and soil - like deposits and litter functioning as a housing medium for microarthropods. The larger epiphytes collect water in which soil and litter accumulate. The numerically abundant taxa are the ants followed by the beetles. Besides the need for a clear understanding of canopy - under storey interactions, a closer understanding of community pollination studies would reveal interesting information (Devy and Ganesh, 2003). For instance, increased Co_2 tends to affect the rates of herbivory which tend to indirectly affect forest productivity. It was Erwin's (1982) investigation on canopy insect diversity that laid a strong foundation for canopy studies. All the same the diversity of technologies now available for canopy studies should evoke increasing interest on the subject in India, more especially when the rainforest canopy is threatened seriously by deforestation.

Diverse technologies like permanent towers, single rope techniques, aerial walking, trunk accessories with both, tree platform etc., have been in use. Ground based methods such as insecticide fogging has enabled researchers to prudently work on tree top (Nadkarni, 1994). Structural and functional aspects of canopy biology are emerging aspect and with both the Western Ghats and N.E. India region being recognized as hot spots, the need for canopy research cannot be over emphasised, providing an excellent input to insect-plant interactions at the canopy and understorey levels, enabling recognition of variation in relation to structure composition and ecological function. Canopy dynamics has therefore emerged as a new dimension in the study of ecosystems. For a better appreciation of the biodiversity in the canopy, some of the well known efforts in this regard available in literature,are presented so as to form a base for future research in this area in India.

The main canopy of the forest is formed by the crown of the middle strata of trees, generally those from 100-130ft. Below the canopy are the shorter trees 50-80ft high, often called the understorey, which consists of shrubs, non-woody plants and young trees. Superimposed in this framework are climbers and epiphytes. The lowest branches of the canopy are around 70 ft or more above ground. The large number of epiphytes survive by extracting nutrients form the dead leaves and twigs that fall from the canopy. Orchids are among the more common epiphytes, besides climbing plants or vines, mostly woody, wavy, flat or curling and their dry loose bank harbour a variety of mycophagous thrips.

Many insects have shapes and colours that offer effective camouflage in their own habitat. Some preying mantises are shaped and coloured like tubular pink flowers and mimicicing by butterflies. As habitat destruction continues to threaten forestry in tropical and temperate forests, canopy will become more reduced and fragmented thereby eliminating many species. The complex interaction between canopy foliage and insects represent a potential side for ecological change, with the alteration of forest ecosystems by man. Needless to emphasise that the complex interaction between canopy foliage and herbivores represent a potential site for ecological change, as human activities continue to alter forest ecosystem (Connell and Lowman, 1989).

Besides the patterns of spatial distribution faunal similarity and distance between trees, the need to understand areas of high biodiversity for priority protection is also important. Microhabitat affiliation, taxonomic and feeding group specializations and body size, enable a better appreciation of tree-crown insect samples. For example, tree crown beetle assemblage involves taxonomic clumping including several families Elateridae, Tenebronidae, Chrysomelidae, Pselaphidae to mention a few (Stork, 1991). In leguminous trees in the Indian context, Bruchidae and Apionidae tend to be associated. A canopy life implies behvioural and morphological preadaption which give rise thro' evolution to more and more adaptable forms, particularly in term of mobility and ability to stay in the canopy, involving mechanism of attachment and modes of oviposition. While grasshoppers include apterous and brachypterous species populations with reduced mobility, the only group presenting important morphological adaptations to canopy life are the *Proctolabinae* which are intracanopy species, while *Romaleidae* is a basic component of canopy life including larger species (Stork, Adis and Didham, 1997). More intense sunlight, extremes of relative humidity, higher water and a better pool of nutrients than in the forest floor are very typical of the canopy. Most epiphytes occur in moist areas much of the crowded appearance being due to mosses, lichens and orchids growing on the trunks and branches of trees surviving in the water and nutrients available. Many ants such as the leaf cutter ants cultivate a special fungus, enabling cultivation of ant gardens. However, ascending the canopy is no easy taks, since one has to overcome stinging ants, rotten trunks and thorn. Several rare and new species of arthropods occur within the tangled vines, under rotten bark and in slimes and moulds (Lasky, 1997). One important aspect is the abundance of detritivores fauna it supports. Species of Guttiferae, Araliaceae, Fagaceae, Moraceae, Sapotaceae, have large accumulations of epiphytes and contiguous mass of crown humus (Nadkarni and Lowman, 1991).

The vertical stratification of arthropod biodiversity is typical of tropical rain forests, the spatial and temporal partitioning or arthropod biodiversity being influenced by host specialization or feeding guilds. As such, alpha diversity within sample units and horizontal and vertical betadiversity are very typical. The epiphytes build up a mat of humus on tree trunks that act as a foraging ground for canopy dwellers, providing resources for a range of herbivores insects such as leaf miners, pollinators, flower feeding caterpillars, the flowers offering nectar and fragrance for bees, flies and butterflies. Bromeliads abound in harbouring a large contingent of mosquito species midges and syrphids. Canopy standing litter involving the host tree epiphytic twigs and leaves resting on branches and trunks abound in mites, Coleoptera, insects larvae, adult ants, Collembola and beetles (Moran, 82). Canopy ants are heterogeneously coloured, others like aphids, caterpillars, grasshoppers show optimum camouflage. Ecological aspects of colouration and camouflage for insect survival in the canopy needs further investigation. Ants which act as predators or scavengers often account for 20-40% of the arthropod biomass of the canopy. In the tropical canopies ants, through fewer in species number, generally show numerical abundance, but beetles are richer in the number of species, though with relatively fewer individuals. Foliage of different species varies in their chemical composition and hence in the diversity of species they may support, so that the diversity of species of plants may mediate the coexistence of competing insects. Densities of all species in tree holes are higher in the canopy layer, in new of the quality as food in higher within holes (Kitching, 1971).

Lichens and other epiphytes growing on tree trunks, branches and leaves are capable of fixing significant amounts of nitrogen as they also capture the minerals carried by rain water and cup fungi play a crucial role in rainforest nutrient cycles by rapidly breaking down tough, woody material. Another basic component relates to the epiphytic mosses and lichens carpeting the upper tree branches. Each species opens up additional means of using the environment for shelter as well as foraging. The ant garden epiphytes or myrmecodia develop tube-like lower parts containing chambers inhabiting ants. Epiphyte rich tropical forests have also such canopy based as genera *Anthurium, Rhododendron, Peperomia,* aroids and bromelids, often called `litter basket epiphytes', which build up an `epiphytic mat' along tree branches, providing, hiding and breeding sites for insects, the mats being a source of moisture and nutrients. Forest canopies therefore have become important in maintaining the diversity, resiliency and functioning of forests since epiphytes contribute to ecosystem processes such as nutrient cycling. At the upper level of the canopy with more sunlight, aquatic breeding habitats for dragonflies, are also known, so that dragonflies feeding in rain forests tend to move upwards. Elsewhere, Anisoptera and libellulids are known to perch on high trees (Paulson, 2005).

Patterns of guild structure are very typical of canopies involving xylophages, herbivores, saprophages, predators and fungivores. Guilds are taxonomically different, but ecologically identical species with many of them being pollination and biocontrol agents. Variability in the proportional abundance of different guilds exists with an abundance of herbivores and fungivores. Guild proportions tend to vary with a high degree of inter tree variation, as exemplified by information on guild structure in Dipterocarpaceae, Anarcadiaceae and Fagaceae. ``Tourist guild'' (Moran and Jonathon, 1982) is defined as `those species having no lasting or intimate association with the plant, but which may be subject to local predators and so form part of the trophic web of the arboreal community''. Dipteran guilds often comprise the following :-

Mycetophilidae Sciaridae	fungal guild
Cecidomyiidae Tephritidae	phytophagous guild
Empididae Dolichopodidae Ceratopogonidae	predatory guild
Tachinidae Pipunculidae	parasitic guild

It is the ephemeral nature and non-specificity of interactious that make them tourists. The role of adult Diptera is questionable, but the length of their larval life and the importance of the larvae as predators, scavengers, fungal feeders and the larval performance in different habitats become relevant. The term ``tourist guild'' has therefore not been applied by Didham (1997) who points out that ``many adult dipteran species are integral components of canopy communities significantly influencing predators-prey and competetive interactions'.

Several guilds are recognized involving phytophages like chewers and sap suckers, epiphyte fauna, scavengers, predators, parasites, ants (Moran and Southwood, 1982). The guild patterns vary with combination of tree species, such as broad and narrow leaved trees or phytophagous, epiphyte feeding and predating guilds. Chewing insects consume 7 to 10% area of tropical forest in canopies (Wint, 1983). Phytophages especially sap-suckers are overwhelmingly dominant, the broad leaved trees tending to support good populations of phytophagous and predatory species. Though insects in hidden cracks, under-bark, leaf-miners and rollers are often overlooked, estimation of the number of individuals in major insect orders, besides species level information are available only for a few groups of insects (though not in India). For instance, Wolda's (1979) work on Homoptera, Erwin's (1982) on beetles in the Amazon forests diversity of the soil fauna of the canopy of a Venezualan forest (Paoletti *et al.,* 1991), and canopy leaf litter ants in a Neotropical forest (Longino and Nadkarni, 1990) offer valuable information. Sufficient variations have also been shown the exist in the number of Homoptera caught between trees, Wolda (1979) showing the clear coordination existing between the number of vines present in the canopy and number of Homoptera. Some of the species include those of membracids such as *Enchenopa notata, Gerridium scutellum,* besides cicadellids, cercopids and psyllids.

Arboreal ants exhibit vertical stratification, some being restricted to greater heights and others to the understorey. Large amounts of leaf litter and other organic material accumulate in the canopy. Arboreal ants like *Camponotus* are stem-nesting ants, while *Solenopsis, Pheidole, Hypoponera, Brachymyrmex* are canopy ants. Stratification is known to exist in the ground and canopy communities and a better understanding of the patterns of tropical ant distribution in the canopy is needed, with respect to microhabitat specialization, besides the an understanding of the vertical stratification of species.

Pollination studies at the canopy level have been restricted to the Dipterocarps. In a notable piece of work, on pollination system, Devy and Davidar (2003) recorded over 50 speceis of anthophilous insects notably social bees, beetles, months, bugs, wasps, thrips and flies from canopy and understory of trees. *Elaeocarpus tuberculatus* pollinated by moths tends to exhibit noctural anthesis, while in other species, anthesis was diurnal. The dipterocaps that dominate the Malayan forests have high species diversity and the dipterocarps guild pollina-tedby thrips is the most dominant (Chan and Appannah 1981, 1990). Thrips known for their high fecundity, increase their population during outburst of mass flowering. Appannah (1979) through examination of 160 flowers of *Shorea lepidotes* over a period of ten days showed that 97.7% of the total number of visitors were thrips, with 469 adults and 10 larvae 166 pollen grains were counted from 79 thrips sampled. A single burst of massive bloom appears typical of dipterocorps, often called ``big bang flowerers", since dipterocarps present their crown above the canopy a feature which suits wind drafts of thrips.

Fig. 5.2. Dipterocarp canopies

Canopy research therefore is a multidisciplinary one involving various aspects of insect-plant interactions, such as photosynthesis, respiration and nutrient cycling and influence of the canopy on microclimate, which in turn have an important impact on the insect fauna. Needless to emphasize that canopy research has come to be realized as an interdisciplinary effort involving various aspects of insect-plant interactions. An understanding of differences in species complexes of canopies, understorey and ground would enable better appreciation of the complexities involved in canopy science. The need for assessment of arboreal communities, notably site-specific aspects, become important more in view of changes in microclimate and associated canopy vegetation, so as to enable a better insight as to how biotic and abiotic factors are linked to canopy structure.

As to whether host specificity exists in the canopy arthropods in terms of the distribution of insects among different host plants, both intrinsic and extrinsic factors tend to restrict their distribution on the host plants. Tourist species which use the plants for mating purposes are less host specific. Canopy fogging samples by Erwin have shown that on an average ``82% of the species were restricted to a single tree species, 12% on two species and 6% on 3 or 4 tree species (Mandsley and Stork, 1997).

CHAPTER 6

HABITAT HETEROGENITY IN AGROECOSYSTEMS : INSECT BIODIVERSITY IN HABITAT EDGES OR ECOTONES

The exploitation of natural resources at a fast rate involving initiation of large scale developmental schemes and multipurpose projects and different types of land use and new farming systems and other agroforestry practices have considerably altered or disrupted the environment of local insect faunal complexes with consequent change in their composition. There can be no better instances then the influence of the Rajasthan Canal on the increased incidence of crop pests and insect vectors as against the decreased incidence of the dreaded red hairy caterpillar *Amsacta albistriga* in Tamil Nadu with the advent of the Parambikulam-Aliyar dam. Tropical deforestation could bring about essentially different ecosystems through its impact on local climate. Human influence and small climatic variation can cause substantial changes in vegetation. The biodiversity for any given landscape will vary substantially over any significant period. A combination of ecological and climatic processes result in a variety of genera, species and ecosystems. The introduction of increased mono-culture has also seriously interfered with insect faunal complexes, not to mention of land use changes and resultant increased changes in species composition, diversity, numbers and biomass of insects. Impact of Co2 on a wide range of insects is an aspect deserving more attention.

Agroecosystems as a whole present a dynamic picture of both crop plants, seeds and other components particularly in view of microclimatic changes. Agrolandscapes are shifting mosaics of crops, pastures, fallow land and woody areas. For instance, throughout the growing seasons, the composition of the crop changes in terms of ground cover, volume and maturity, so that considerable movement of insect species occur such major pests as *Nilaparvatha lugens,* the brown plant hopper and *Sogatella furcifera*, the plant hopper of rice, which migrate long distances every year occupying favourable habitats causing considerable damage to the rice crop. The development of insect biotypes has posed a serious threat to host plant resistance and the continuous growing of insect resistant cultivars tend to lead to some behavioural and physiological changes in these insect pests, so that they become capable of feeding and developing on resistant varieties. In short, biotypes refer to populations within a species with the capability of surviving and destroying varieties that have genes for resistance. Besides *Nilaparvata* and *Sogatella,* species which show biotypes are *Nephotettix virescens, Orseola oryzae* the gall midge of rice, *Rhopalosiphum maidis,* the corn leaf aphid and *Schizaphis gramimum* the sorghum and wheat aphid.

Impact of climate changes in agroecosystems

Decline in the nutrient quality of plants tend to result in poor performance of insects with the tendency to become less abundant as atmospheric Co_2 levels increase. Aphids being important pests of agriculture and horticulture offer unique opportunities for studies on interaction with climatic changes. Their rapid and dramatic responses to change, in particular to temperature is aided by their short generation cycles and immense capacity for reproduction. Other variables like Co_2 and pollutants also tend to have a direct or indirect effect on aphid population. Further, the effects of such climatic changes also tend to have their own impact on other host plants, natural enemies, competition making their interactions complex. Life cycle variation in *Myzus persicae* around the world depend on interactions climate and the genotype. For instance, the potential for sexual reproduction may be retained throughout the range of the species, but suppressed in some areas due to the absence of photoperiodic cues necessary to trigger it or to high temperature which inhibits photoperiodic responses. Besides photoperiod and temperature, moisture is an important cue for the seasonal behaviour of insects.

Intensification of agriculture or Modernisation of agriculture

While industry is largely responsible, intensification of agriculture and animal husbandry, has also increased the rates of emission of gaseous volatiles such as methane and ammonia. Annual methane releasing rate at a global level indicates that natural wetlands, rice fields, biomass burning, landfills and gas drilling contribute to nearly 95%. Such volatile organics have serious implications in relation to plant growth, which indirectly affect a host of insects. Lifecycles of annual insects are interfered by SO_2 and all the three pollutants SO_2, NO_2 and ozone are known to have positive effects on insects of agricultural, horticultural and silvicultural importance, with many of them being prone to rapid responses with immense capacity for reproduction. Besides, changes in rainfall patterns have their impact, in particular of greater soil erosion and new crops and varieties will have to adapt to a greater degree of soil and water stress. Contamination of agricultural land through long term use of pesticides and increased fertilizer input, besides the introduction of new crops or cultivators contribute to changes in pest and natural enemy complexes. Following insecticide application there is usually an abnormal increase in pest population, mostly Homoptera (41%) followed by Lepidoptera (24%) and phytophagous mites (26%). Resurgence may occur due to a variety of factor related to the host plant, the insect pests, non-target organisms in the agroecosystem and the insecticide used. Chelliah and Bharathi (1993) have devised a `pest resurgence chart identifying host of factors influencing pesticide induced resurgence of insect pests.

Ecological consequences of modern agriculture

Expansion of irrigation facilities has also caused an aggregation of pest problems. Water logging enhances multiplication of several borer pests like stalk borer, internode borer and white fly in the sugarcane crop. Frequent irrigation also causes increase of *Pyrilla perpusilla* populations as also increase in aphids. Many other practices like planting time, growing intercrops, trap crops, trenching else tend to contribute to aggravation of insect pest populations. (Dhaliwal and Arora 1996).

Ecological consequences of modern agriculture with special reference to pest problems, have to the borne in mind while intending to achieve maximum productivity per unit time from available land. Though such practices as spacing, crop rotation, sowing time, tillage methods contribute to increased yields, they have also contributed to severe outbreaks of insect parts. With high yielding varieties being introduced in crops like rice and wheat, several new pests have been recorded, besides minor pests achieving a major status. As has been indicated earlicr, insects like the brown plant hopper, white backed plant hopper, green leaf hopper, gall midge, leaf folder and whorl maggot have become major pests of rice since the introduction of high yielding varieties (Dhaliwal & Arora, 1996). As such modern plant breeding with its emphasis on high yielding varieties depletes the genetic base of many crops. Resistant varieties have been developed against a number of pests of rice (gall midge, stem borer, brown plant hopper and green leaf hopper), maize (maize stem borer and pink stem borer); sorghum (shoot fly and midge); ground nut (*Spodoptera,* leaf miner, jassids and thrips), cotton (bollworms and jassids); sugarcane (top borer, scale insects, mealy bugs and white fly) and pea (pod borer and leaf miner) (Dhaliwal and Arora, 1996).

Improvement in technology notably in the case of unploughed land, has resulted in the loss of wild populations, besides the variety of long established seminatural vegetation, so that chances for recolonization become sparse. Besides, there is the almost impossible aspect of restoring soil profiles as well as uncommon vegetation which harbour diverse insects pollinators and predators. Such efforts enable sustainance of ecologically important species which are essential for the effective functioning of agroecosystems (Corbet, 1995).

Indiscriminate transport of sugarcane setts has enabled insect pests like borers and scales to new areas large in acreage. The continuity of the standing crop in some stage or other throughout the year is largely responsible for the severity of insect pest problems in sugarcane and more than 200 species of insects are reported from this country. Further continuous monoculture of sugarcane, have resulted in the multiplication of insect pests (Dhaliwal and Arora, 1996). Maintaining or ever increasing the diversity and number of pollinating insects are linked with the landscape. Agriculture landscape should be shaped in such a way as to make sure that the arable land covered does not exceed 75% of the total. Bee populations should be augmented as also pollination practices, since better pollination of arable crops means increased food for farmers. As such, for a proper management of agroecosystems, an ecological perspective of the effects of habitat heterogeneity tends to act as a powerful tool controlling pest populations by manipulating vegetation diversity in agroecosystems. The intensification of agriculture utilizing high yielding varieties grown under new agronomic practices with increased irrigation facilities and extensive use of agrochemicals have resulted in increasing incidence and outbreaks of several pests over the last few years.

Trap crops

The planting of trap crops next to higher value crops, is more attractive to a pest as either a food source or oviposition site than a main crop, preventing or delaying the arrival of pests to the main crop. Trap crops act as plant stands that are *per se* or via manipulation to attract, divert, intercept and or retain targeted insects or pathogens they vector in order to reduce damage to the main crop (Shelton and Badones - Perez, 2006). It is a habitat manipulation of an agroecosystem for purposes of pest management. Different herbivore insect species feeding on a plant population tend to form guilds which are ``a group of species that exploit the same class of environmental resources in a similar way". In other words, guilds are elementary communities, populations or individuals of a guild influencing each other. In *Brassica oleracea* or cabbage there are the following feeding guilds.

1. Pit feeding guild, consuming small patches of leaves by flea beetles, weevils.;
2. Strip feeding guild, chewing leaves from the sides.
3. Sap-feeding guild as in aphids, leaf hopper, tingids, thrips.

Increasing vegetation diversity by planting different crops intermingled is a cultural control strategy, that tends to make agroecosystems less favourable to pests and more favourable to the natural enemies-such as intercropping, wherely more than one crop is grwn in an area simultaneously. The types of intercropping are known – *mixed cropping* growing two or more crops simultaneously without distinct rows; *row intercropping* with one or more of the crop grown simultaneously in rows; *strip intercropping* with two or more crops grown in strips wide enough to permit independent cultivation. Trap cropping where one species serves as a trap crop, to divert the pest away from major crop (Schoonhoven, *et al.,* 19980. A trap crop planted next to a higher value crop is more attractive on to a pest as either a food source or oviposition site than a main crop, preventing or delaying the arrival of pests to the main crop. Shelton, *et al.* (2001) define trap crops ``as plant stands that are *per se* or via manipulation deployed to attract, divert, intercept and or retain target insects or the pathogens they vector, in order to reduce damage to the main crop". It is a habitat manipulation of an agroecosystem for purposes of pest management. For example, tomato monocultures were totally destroyed by *Spodoptera litura* caterpillars, while in an intercrop of tomato and beans did not, since the caterpillars were attracted more to beans. Trap cropping has proved beneficial cotton, soyabean, potato, cauliflower.

Habitat edges or Ecotones

The boundary separating two different types of habitat, viz `edge habitat' or ecotones harbour several species. With many new

Fig. 6.1. Habitat edges in the rice crop (After Ravi)

Fig. 6.2. Habitat edges with *Echinochloa* and *Brachiaria* weeds (After Ravi)

Fig. 6.3. Rice Hispa on weeds (After Ravi)

Fig. 6.4. Weed *Isilema* with midge galls, ricethrips and mite infestation (After Ravi)

macrohabitats becoming increasingly fragmented, species inhabiting microhabitats gain increasing importance with many of them exhibiting variation in patchy density, so that opportunities to study diversity in these microhabitats become considerable. Urban development leads to several kinds of habitat loss, both qualitative and quantitative. As indicated earlier, four types of habitat loss are known viz. loss of quality, quantity, loss of connectivity and continuity (Hanski, 2005). In recent years many ecosystems have been replaced and small remnants are left in a landscape modified by agricultural or monocultural forests. Though this reduces biodiversity, the existence of `edge effects' due to species dispersal from neighbouring habitats tend to add to biodiversity in a small way. Many edge habitats have diverse communities depending on the type of vegetation and they are vulnerable to loss and becoming reintroduced in course of time. Edge effect exists in such boundaries, as seen in agricultural and forest landscapes, where the boundaries between forests and fields and between forest and clearcut areas are sharp (Fig. 6.1).

Such ecotones with their special fauna also tend to disappear with adverse conditions. Habitat edges like bund or hedgerows in rice, sugarcane ecosystem harbour such species so that a knowledge of such habitats become relevant for a better assessment of the periodicity of invasion by pest species. As such from a single field agroecosystem approach, an agrolandscape approach has emerged in which landscape elements such as habitats edges or ecotones or corridors are included and patterned to involve aspects such as habitat fragmentation, trophic and biotic diversity and degree of connectivity. The dispersal behaviour of insects and natural enemies become essential aspects of such changed ecosystem.

The abundance, diversity or effectiveness of natural enemies of pest may be enhanced by the presence of uncultivated corridors, particularly where flowering plants provide the necessary nectar, pollen and microhabitats. Some of the patterns of field border vegetation on the diversity and abundance of arthropods in crops are : (1) grassland adjacent to crops can significantly increase colonization by predatory arthropods (2) field border vegetation serve as a reservoir for crop pests (Alteri and Letourneau, 1982). Predaceous Hemiptera tend to be more abundant in successional, than grassy corridors. The size and directional of seasonal change in abundance in most taxa different between the corridors (Rodenhouse *et al.,* 1992).

Insects in agrolandscapes are not only restricted to crops fields, but the survival of many species depends on resources outside cultivated areas such as weeds bordering fields, hedgerows and the like. Insects inhabiting such environments find a place in this chapter, notably around the paddy field ecosystem. Management action outside the fields will tend to have increased impact on insects within the field. This is because, pollinators, soil living insects as well as biological control agents depend on weed hosts at the field margins as well as nearly vegetation. Introduction of crop varieties with high nutritive value and restriction of natural enemies through destruction of habitat and pesticide usage are highly detrimental. As such a better appreciation of changing landscapes as well as the movement capacity of insects became relevant. One of the principal actions is to link population dynamics to spatial and temporal habitat quality, for which a better understanding of *connectivity, configuration and composition* become necessary inputs (Ekbom, Erwin and Robert, 2000) (Fig. 6.2). While connectivity refers to the movement of habitats, the proximity of different habitats to each other (configuration) is equally important. Fragmentation of habitats caused mainly by the loss of connectivity and dominance of crop fields in habitat composition have become essential elements in conservation biology. The structure of agrolandscape therefore is a system of patchy habitats, corridors and environmental barriers which form an ecological network within the matric of arable fields (Forman and Gordon, 1986).

Biotic communities in crop ecosystems, notably weed species have established themselves as an important element in insect pest-natural energy interaction, being an important source for the latter. Certain weeds affect the biology and dynamics of natural enemies, in many cases also tending reduce pest species to move away from the crops. A great variety of insect species in grasses enables a continuous flow into cultivated fields, so that the role of grasses and other weeds serve as the original or alternate host of pest species.

Weeds may influence the insects in many ways (Way & Cammel, 1981) :

(A) by acting as alternate or alternative hosts of the pests;

(B) by affecting the biology of insects pests;

(C) by affecting the visual responses of flying insects;

(D) by affecting the behaviour of pests once they have contacted the weed; and

(E) by influencing pest morality, as a result of harbouring natural enemies of insects.

Alternate host plants are particularly important where the cropping season is short and followed by a cold or dry season of fallow or total replacement by a different cropping system (van Emden, 1981). Uncultivated land provides plant debris and other forms of physical shelter for pests hibernating in active state. Insects which overwinter as adults (especially bugs and beetles) nearly all left the crop to hibernate in the shelter of cultivated land, and such insects include the most important pests of crucifers and legumes. Thus, the pest reservoir in uncultivated land is often the source of the initial infestation, and the insects move onto the crops for diverse reasons. One of the famous instances relate to *Limothrips cerealium* in UK which overwinter under bark of trees and with the onset of spring move in a mass to oatfields (Lewis, 19).

Weediness in short, has diverse characteristics such as successful establishment over a wide range of environmental conditions; seeds are long lived with continuous productivity; cross pollination effected by wind or unspecialized insects; high seed production; efficient seed dispersal mechanism; rapid life cycles; efficient vegetative reproduction or regeneration from fragments and are highly competitive for resources. (Rechcigl and Rechcigl, 2000).

Weeds are therefore so prevalent that they are an important component of agroecosystems. The fact that many pest species colonize on weeds, and breed successfully, reveals that they can migrate to the crop and inflict severe damage at any time. This is more so when the growing period of crop plants synchronize with that of the weeds. Instances where there are definite host-plant successions between crops and weeds are also on record. Thus a knowledge of the nature of colonization of insect pests and their population build-up on crop and weed hosts become essential to understand the intersections between insects, crops and weeds.

Host plant Switching

Another aspect with regard to the population pattern of insects on crop and weed hosts is host plant switching. Host plant switching is normally encountered when (1) the crop/weed dries up. (2) the exact stage of the host plant is not available for the Insect to feed, or (3) if there is any nutritional inadequacy so that insects tend to disperse to plants of more nutritive value for feeding and breeding. For example, assessment of the weed hosts of several heteropteran bugs have shown to contribute to the existence of several egg parasitoids, thereby the natural enemy populations are maintained optimally and subsequently migrating to the crop plants when their host populations appear. For example Velayudhan (1987) reported that *Nezara viridula* occuring in more than a dozen cereal crops as well as in weeds such

Croton sp., *Cleome viscosa, Gynandropsis* sp. during major part of the year. Egg parasitoids such as *Acroclisoides* sp., *Anastatus ramakrishnae* and *Telenomus lucellus* parasitize *Nezara viridula* eggs on both crops and weeds. The populations of these natural enemies are maintained throughout the year by parasitizing the eggs oviposited in weeds. Similar interactions have also been reported by Velayudhan *et al.* (1988) and Senrayan *et al.* (1988) for other pentatomid and coreid hosts. Senrayan (1989) also reported the role of the perennial weed in the maintenance of the stink bug predatory population of *Eocanthecona furcellata,* a predator of several lepidopterous larvae. The populations of *E. furcellata* primarily appear in *Mangifera indica* destroying caterpillars of *Latoia lepida.* Adjacent weeds such as *Cassia marginata* also support this predator population. Therefore, weed hosts not only favour the herbivores by acting as reservoirs, but also provide a sort of a refuge to naural enemy population to have their continued interaction with pest population. *Vicia sativa* is a common weed associated with chick pea in North India. Removal of the weed at a time when maximum eggs one laid, substantially reduces the incidence of *Helicoverpa armigera* (Dhaliwal and Arora, 1996).

Over the last three or four decades increased manipulation of the landscape for agricultural production has resulted in the loss of natural habitats. Natural corridors have been an important component of the landscape. Human established corridors fence rows, hedgrows, road-side vegetation have been essential elements of fragmented landscapes. The corridors could be (a) planted corridors or shelter belts in agricultural landscapes to prevent soil erosion and enhance biodiverity (b) regenerated corridors, which are strips of vegetation along road sides (c) remnant corridors which are strips of natural vegetation. All the three provide niches for insects and biocontrol agents.

Weedy species are typically characterized by some of the following features (a) special adaptations for long and short distance dispersal (b) rapid growth at initial stages (c) tolerance to climatic vegetation (d) great competitive ability (e) early reproductive capability (f) have several generations (Watt *et al.,* 1997). As such uncultivated corridors should be established within cropland to reap multiple benefits. The presence of specific weeds in habitats adjacent to crop fields determine to a considerable extent the type of predatory species.

Weeds therefore have a major role to play in the harbouring of insect populations. Well known examples of such insects are referable to vegetation in bunds of rice, sugarcane crops or in bordering vegetation. Insect conservation biologists look to connectivity and permeability as the two characteristics of hedgerows. As weeds are a major component of agroecosystems affecting the biology of pests and natural enemies, their role in regulating pest populations become important. Being major components of agroecosystems they make a definite contribution to the richness of the fauna within these systems. The type of weed community and associated entomofauna are useful in increasing community and insect stability in the area. Weeds are frequently the only source of pollen and nectar in flowers, which are vital in maintaining high population of beneficial insects within an agroecosystem (Figs. 6.3 and 6.4).

Therefore an increasingly popular mode of habitat diversification is the incorportion of natural vegetation in and around crop areas in an attempt to reduce pest populations. This supports the `resource concentration hypothesis' of Root (1973) which stipulates that more diverse habitats tend to have lower pest densities, because increased ones inhibit the ability of `herbivores to find host plants". Secondly the `natural enemy hypothesis' of Root implies that diverse habitats tend to harbour more predators affording more protection against pests.

Some common weeds and associated insects

Species	Weeds	Crop
Oxya nitidula	*Graminaceous and Cyperaceous hosts*	Rice
Cyrtacanthacris ranacea	*Abutilon indicum* *Sida rhombifolia*	*Gossypuim* *Abelomoschus*
Orthacris maindroni *Atractomorpha crenulata* *Truxalis indicus*	*Graminae*	Rice
Chirothrips mexicanus	*Chloris barbata*	*Pennisetum typhoides*
Haplothrips ganglbaueri	*Echinochloa crusgalli*	Rice
Stenchaetothrips oryzae	*Echinochloa colona*	Rice
Caliothrips indicus	*Achyranthus aspera*	groundnut
Zaniothrips ricini	*Datura stramoneum* *Calotropis gigantea*	Castor
Megalurothrips indicus	*Andrographis* sp.	*Arachis hypogea* *Phaseolus aureus*
Mylabris pustulata	*Ipomea hederacea*	Cotton
M. balteata	*Tephrosia purpurea*	*Curcurbita maxima*

Cyrtopeltis tenuis	*Cleome viscosa* *Gynandropsis pentapylla*	Tomato *Luffa*
Spilostethus pandurus	*Calotropis gigantea* *Vernonia cinera*	Sesamum Gossypium
Oxycarenus hyalipennis	*Abutilon indicum* *Sida rhombifolia*	Cotton

Conclusion

Establishment and maintenance of more suitble habitats on farm or in the surrounding landscape can enhance the survival of both introduced and naturally occurring parasitoids, there improving biological control. Both endemic and weedy non-crop species across a highly modified agrolandscape support species rich and abundant parasitic wsp assemblages with diverse biology and host association (Stephen, *et al.,* 2006). *Trichogramma, Telenomus, Ceranisus* species are common and important biocontrol agents for lepidopteran, hemipteran and thrips pests. Weedy plant species tend to provide an important reservoir of mobile parasitic wasps of benefit to crop plants grown in adjacent fields influence parasitic diversity. So increased habitat complexity tends enhance natural parasitoid population. Increasing natural vegetation in and around crop fields alters the ability of insects to move through the habitat and this dispersal ability can greatly influence the distribution of population of prey and predatory species. As such ecological aspects of habitat heterogeneity is emerging as a powerful tool for managing agroecosystem. Growing of mixed crop, using mixtures of a crop and natural weedy vegetation tend to be beneficial.

Agroecosystems therefore when viewed as a whole presents a picture of dynamism of both crop plants and weeds as well as of insects infesting than ploughing the field at different times of the year affects the species composition and abundance not only of the weed complex, but also of insect species. Such a seasonal disturbance of weeds also results in changes of the predatory complex which fluctuates according to the availability of alternate prey (Altieri and Whitcomb, 1978/79). The presence of specific weeds in habitats adjacent to crop fields determines to a considerable extent, the type of predatory species inhabiting particular crop fields. As such the interaction of predator, prey and host plants may differ from one crop to another from year to year. Manipulation of plant diversity in agroecosystem through seasonal disturbance of soil by ploughing is being in vogue in many areas.

CHAPTER 7

LITTER ECOSYSTEM DYNAMICS IN NATURAL AND DISTURBED FOREST ECOSYSTEMS

Human interference with forest ecosystem in general and tropical rainforests in particular has caused widespread concern regarding their consequences which have been known to affect climate, biodiversity and human societies. While biomass is created by photosynthesis in virgin stands, the process is reversed during forest denudation with the loss of litter resulting in poor soil formation, reduction and loss of biodiversity. Extensive denudation affects biological control resources, some of which at low densities would be vulnerable to genetic erosion or change extinction, with their reduced numbers. The litter ecosystem provides a typical instance of a metabiotic process wherein one series of organisms provides very favourable conditions for the next. With human interference resulting in the depletion of both the quality and quantity of leaf litter, the trophic chain gets depleted. The role of forests in establishing microclimate is well known and the soil cover provided by the litter harbours a wealth of insect resources. Therefore the preservation and protection of forests to safeguard natural ecosystems as well as the genetic pool they contain are important aspects which should deserve consideration, particularly as regards the active role of litter organisms in the breakdown of litter as well as sustaining the food web or trophic complexities in different forest types both disturbed and natural, to merit comparison. The natural forest ecosystem with its floral diversity and specialized microniches, supports a variety of micro-and macroarthropods involving different trophic levels. A knowledge of food web structure becomes very relevant involving categories like decomposers, detritus feeders, mycophages, polyphages and predators (Fig. 7.1).

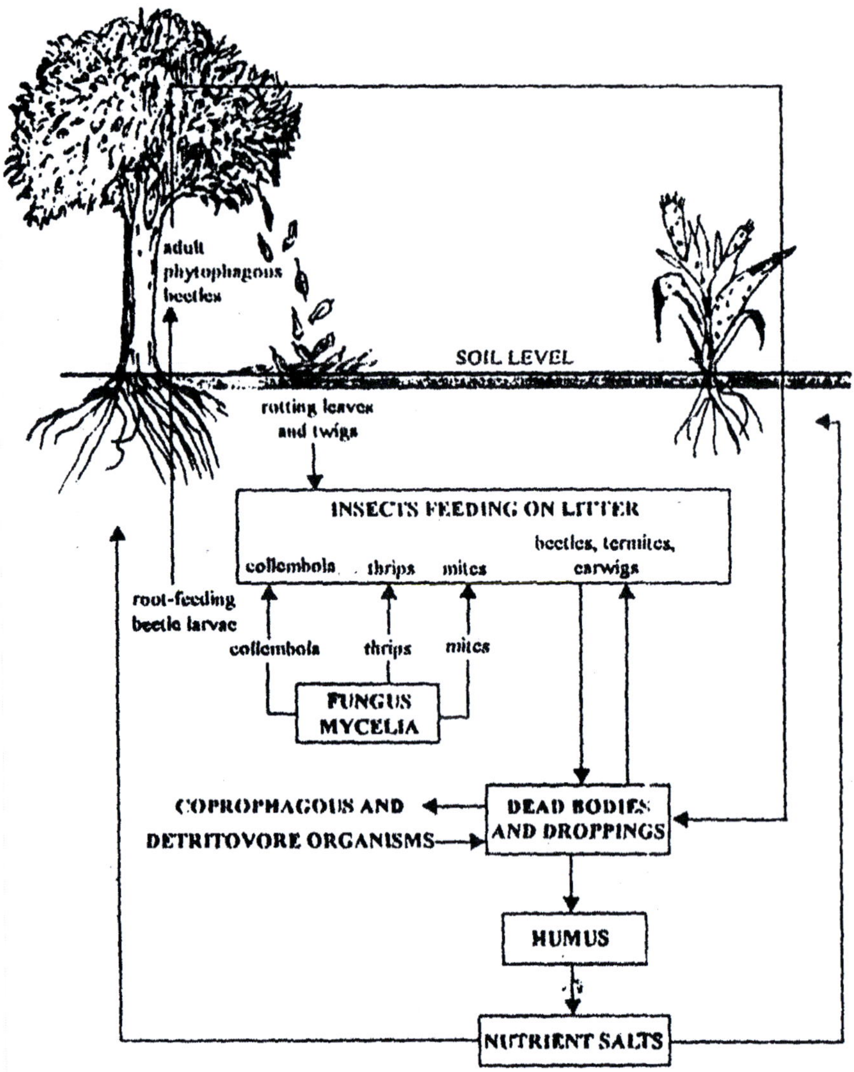

Fig. 7.1. Schematic diagram showing some food relationships and soil Organisms.

The biotic resource of the litter ecosystem involving diverse insects and microarthropods, as well as parasite and predatory species are highly variable in exploited forests and undisturbed natural forests. Certain species of microarthropods have been used as indicator species for particular stages of decomposition such as the *bacterial phase,* succeeding the acari phase and the *humification* phase involving Collembola, thrips, mites etc. Several species of fungi in the litter ecosystem support quite a number of mycophagous species. As many as 32 species of fungi have been recorded on teak litter. Further, the chemical nature of the litter involving inorganic components as well as the phenolics, flavonoids and volatiles also exert considerable influence on the degree of complexity of the litter community. Based on the quality of the litter layer, colonizing insect species vary in the different layers. Analysis of the vertical components of the litter elements has shown that the topmost layer comprising the predatory chain harbour groups like Coleoptera Orthoptera and Hemiptera, the intermediate layer with partially decomposed litter with thrips earwigs Collembola and the third or humus layer mainly composed of Collembolans and mites. Interestingly enough in natural polyculture forest litter collembolan dominate, while mites dominate in the monoculture forest litter.

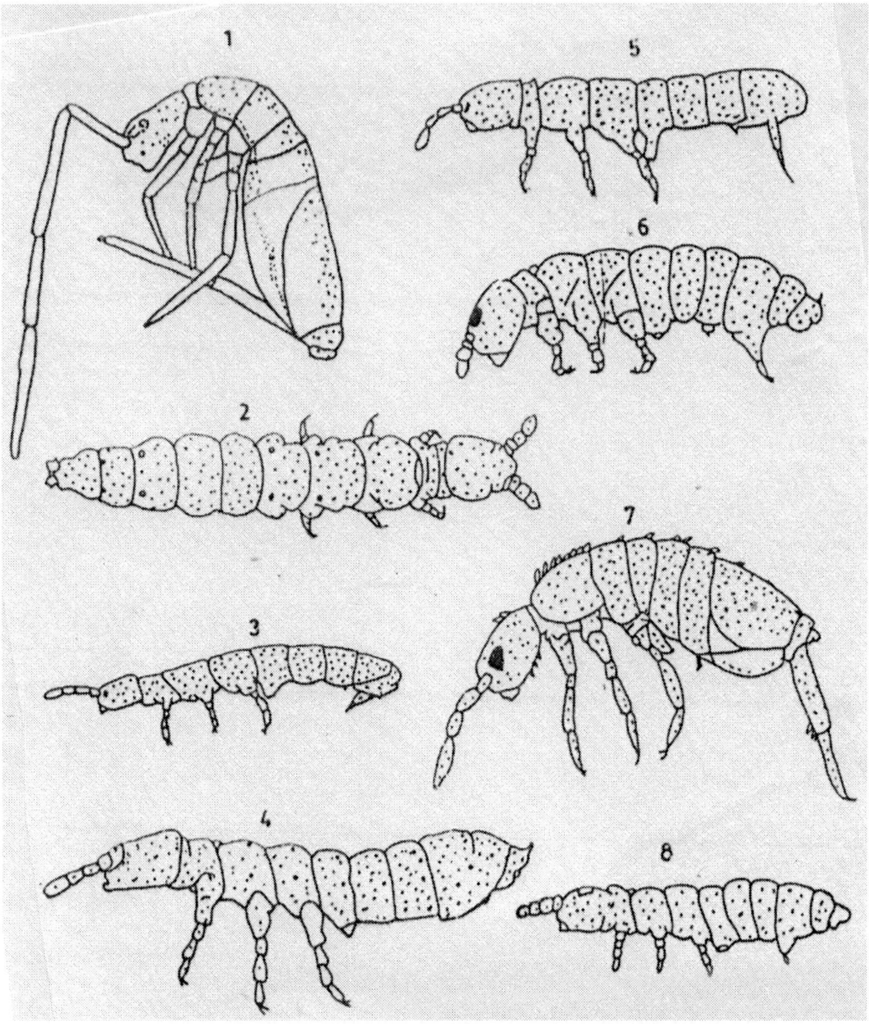

Fig. 7.2. Some common litter Collembola
1. Salina; 2. Tullbergia; 3. Isotomodes; 4. Onychiurus; 5. Folsomia; 6. Hypogastrura; 7. Lepidocyrtus; 8. Xenylla.

Among the litter arthropods in general, Collembola dominates with mites ranking next. Onychiuridae, Isotomidae, Poduridae and Entomobrydae are richly represented in the litter both in number and species composition. Larger forms like *Tomocerus, Entomobrya, Orchesella* abound in surface litter while smaller form like *Onychiurus, Tullbergia* occur in the deeper layers (Fig. 7.2). Surface litter is inhabited by larger mites such as *Macrocheles, Pergemasus, Steganocarus* and smaller ones such as Scheloribates and Brachythomus which penetrate into the humus. The vertical stratification of the collembolan and the other entomofauna could be due to the vertical stratification of the litter horizon, the nature of the litter and microclimatic adaptation of the colonizing species.

A comparative assessment of the entomofauna in three different monoculture litter habitats eucalyptus, tea and coffee showed considerable variation both in terms of species diversity and density. Eucalyptus presented a more complex pattern with acarine population always dominant (91-180/m^2), while tea litter was devoid of thrips, scarabaeids with low population of collembolans and acarines. The coffee litter was devoid of carabids, earwigs and thrips, with low population of acarines. The individual species of Collembola in the three different litter zones - forest lieaf litter, middle and humus zones showed that in the top layer *Callyntura ceylonica, Salina bicolor, Salina quntturofasciata,* the middle layer with species of *Xenylla, Hypogastrura, Cryptophagus and Tullbergia.* The humus zone comprised all species of the middle zone with *Xenylla* dominating followed by *Hypogastrura, Callyntura, Tullbergia and Folsomoides.* A considerable difference in temperature at the different zones was evident 21-23°C in the top player, 20-21 and 19-21°C in the middle and humus zones (Annadurai, *et al.,* 1988). The surface layer of litter abounds in macropredators such as reduviids, anthocorids and Lygaeids, Coleoptera such as carabids, staphynids, pselaphids, besides earwigs and chelonethi, predatory collembolans, mites and thrips comprise the micropredators. Some primitive mycophagous Terebrantion thrips are rare litter inhabitants.

Ants tend to exert a strong structural influence on arthropod communities of tropical rainforest and hence an understanding of ant assemblages become important. Species of Formicinae, Myrmicinae, Dolichoderinae and Ponerinae are strongly represented, the myrmicinae with genera such as *Pheidole, Strumigenys, Tetramorium, Monomorium, Crematogaster, Solenopsis and Myrmicaria.* The Dolichoderinae is represented by *Tapinoma* and the Ponerinae by *Sturmigenys, Leptogenys, Pachycondyla* and *Diacamma.* The Formincinae include species of *Camponotus*. More recently forest management agencies in Australia have incorporated ants in monitoring programmes associated with fire, grazing and logging practices. The use of functional groups in relation to environmental stress and disturbance has come into vogue enabling prediction and interpretation of ant community responses to land use (Anderson, 1987; Vanderwoude *et al.*, 1997).

The conspicuous diversity in leaf production and constant non synchronous leaf fall make available abundant litter which provide an ideal environment constituting a saprophytic fungal zone. The major contribution to the litter in forest areas like the Western Ghats involve species of *Messua ferrea* Guttiferae), *Ternstromoeia* (Ternestromiacea), *Dipterocarpus* (Dipterocarpaceae), *Pterospermum* (Sterculiaceae), *Pterocarpus* (leguminoseae), *Terminalia* (Combretaceae), *Barringtonia* (Lecithidacea), *Lagerstromeia* (Lauraceae), *Kigelia* (Bigoniaceae) to mention the major ones. Apart from microarthropods listed earlier the dominant species of litter (Dipterocarpaceae), *Pterospermum* (Sterculiacene), *Pterocarpus* (Leguminoseae), *Terminalia* (Combretaceae), *Barringtonia* (Lecithidaceae), *Lagerstromeia* (Lauraceae), *Kigelia* (Bignoniaceae), to mention the major ones. Apart from microarthropods listed earlier, the dominant species of litter thrips belong to the genera *Elaphrothrips, Hoplandrothrips, Hoplothrips, Azaleothrips, Stigmothrips, Nesothrips and Urothripids.*

Table 1. Feeding guild wise distribution of ants

S.N.	Predators	%	Prey preference	S.N.	Omnivores	%
1	*Myrmicaria* sp	0.21	General	1	*Camponotus angusticolis*	0.64
2	*Myrmicaria brunnea*	0.21	General	2	*Camponotus radiatus*	0.42
3	*Crematagaster* sp	0.42	General	3	*Camponotus* sp 3	0.21
4	*Solenopsis* sp 1	1.28	Termite	4	*Monomorium* sp	0.21
5	*Solenopsis* sp 2	15.02	Termite	5	*Tetramorium* sp	0.21
6	*Strumigenys* sp	21.88	Collembola	6	*Tapinoma*	51.07
7	*Leptogenis* sp	7.29	Termite			
8	*Pachycondyla* sp	0.42	Termite			
9	*Diacamma* sp	0.42	General			
	Total	**47.15**			**Total**	**52.76**

(after Sabu Thomas)

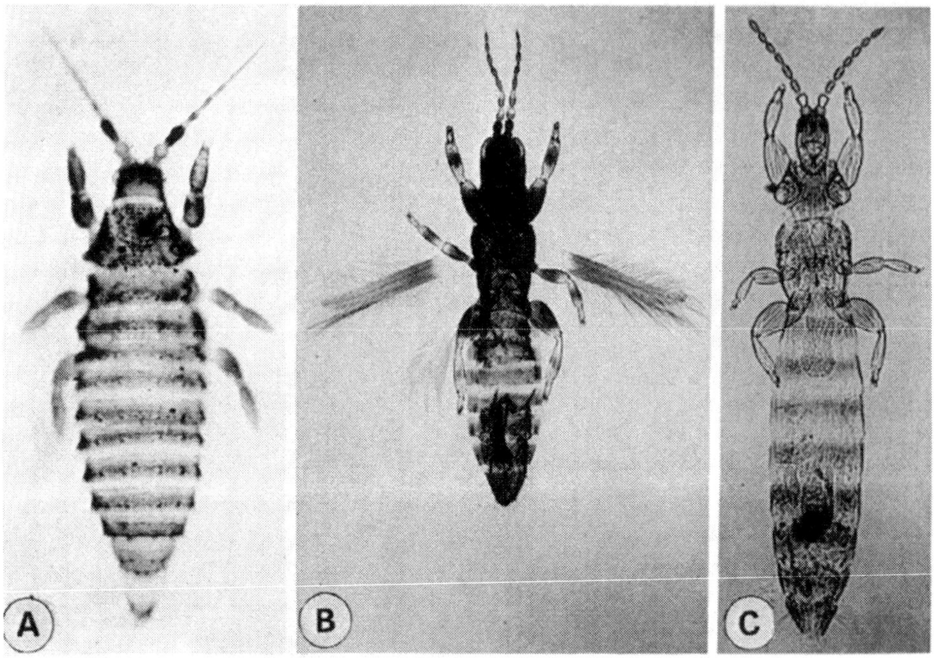

Fig. 7.3. Primitive mycophagous thrips in litter

Shifting cultivation and entomofaunal diversity

Shifting cultivation, a land use pattern practiced by tribals of North East India causes large scale damage, such as loss of diverse vegetation, degradation of soil fertility and consequently decline in the productivity of tropical rain forests. Repeated burning and cropping lead to a decline of nutrients and organic matter in the soil. Comparative assessment of the litter fauna including diverse predatory entomo resources in fallow sites such as after 1, 5, 10 and 20 years, showed wide fluctuation in faunal diversity and density in relation to the length of the fallow period. During the process of slashing, burning and cropping there tends to be a complete disruption in the population density and species diversity.

Comparison of the entomofaunal resources associated with shifting cultivation areas, reveals their abundance in late fallow and natural forest litter compared to early fallow sites. Early fallow areas supported thin vegetation with a dominance of the weed *Lantana camera.* Although this species was available in late fallow areas, a thick cover with varied plant species tends to increase the complexity of the litter wherein microhabitats, feeding choices and abiotic factors provide conducive environment for the establishment and population growth of litter insects and their prey species. Collembolan and mite populations show that their populations were higher in natural forest litter compared to monoculture plantations (Annadurai *et al.,* 1988) Fig. 7.4. This type of faunal diversity could be related to the richness in terms of physical and chemical nature of litter contributing plant species. The diversity of insect communities is directly correlated with the nutritional status of the litter particularly the readily available primary nutrients. The quality of litter biomass is the primary factor regulating the composition of microbes and fauna in any litter habitat. The impact of chemicals released during litter breakdown and shifts in the relative abundance of primary or secondary compounds including the phenolics, besides carbon, nitrogen and water ratios regulate faunal and microbial divesity and abundance.

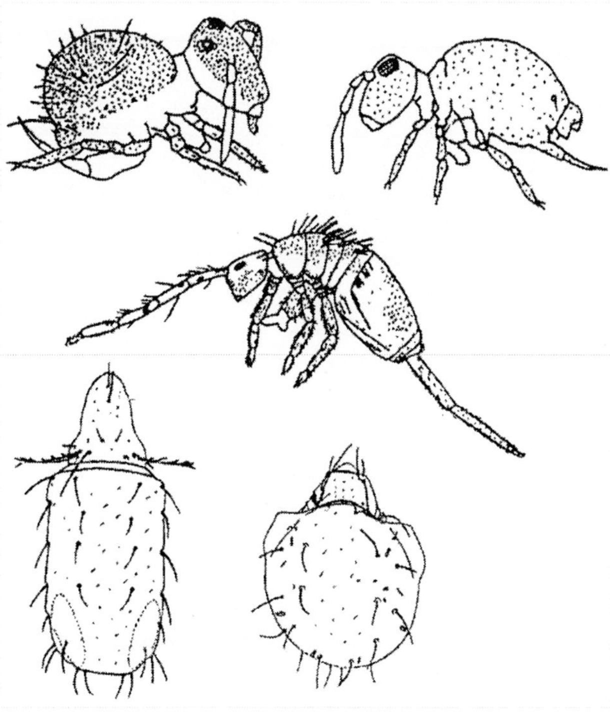

Fig. 7.4. Some soil collembola and mites

The vertical stratification of the types of litter microorganisms suitable for a particular microarthropod species or their immatures may be largely responsible for species distribution pattern in the field. Combination of high species diversity with a comparatively low degree of feeding specificity in detritivore communities, such as Collembola suggests that interspecific competition is possibly reduced not only different vertical distribution, but also by the horizontal compartmentalization of resources. The vertical stratification of the collembolan species and the other entomofauna could be due to the vertical stratification of the litter horizons, the nature of the litter, and the microclimatic adaptations of the colonizing species. The abundance of large collembolan forms in the top litter layer and cryptic forms in the middle and humus zone could also be a behavioural strategy to escape from the predatory arthropods and this differential vertical stratification may be a temporal displacement since many species are not restricted to any one particular layer. Some of the temporal changes in the species composition of populations associated with decomposing litter have been correlated with physical and chemical properties of the litter. Certain microarthropods could be used as indicator species for particular stages in the decomposition of manue. The occurrence of higher population of such collembolan species as *Xenylla sp., Tullbergia sp., and Folsomoides sp.,* is a direct indication of a decomposed litter with a thick humus layer.

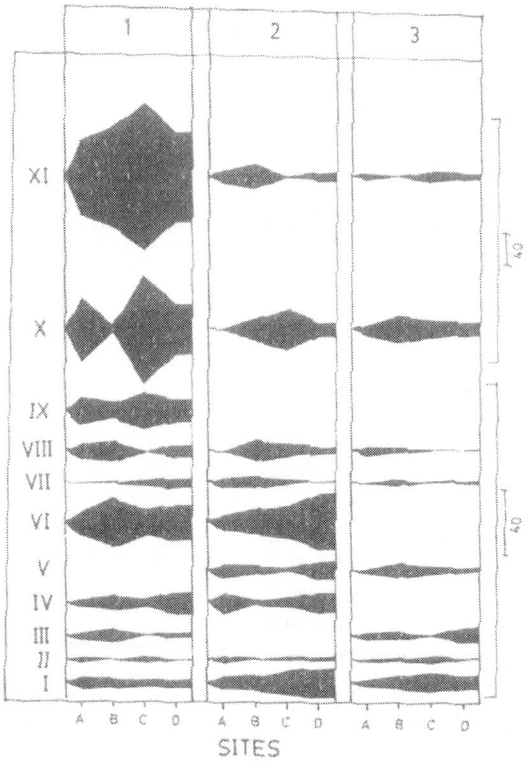

Fig. 7.5. Entomofaunal resources in monoculture litter
1,2,3 Relation composition of microarthropods. 1, natural orest litter; 2, 3. monoculture forest.

Further the trophic chain in a natural litter habitat is more complex than a monoculture litter and the absence of important trophic links in the monoculture litter indicates the loss of potential bioresources in such monoculture habitats as coffee, tea and eucalyptus.

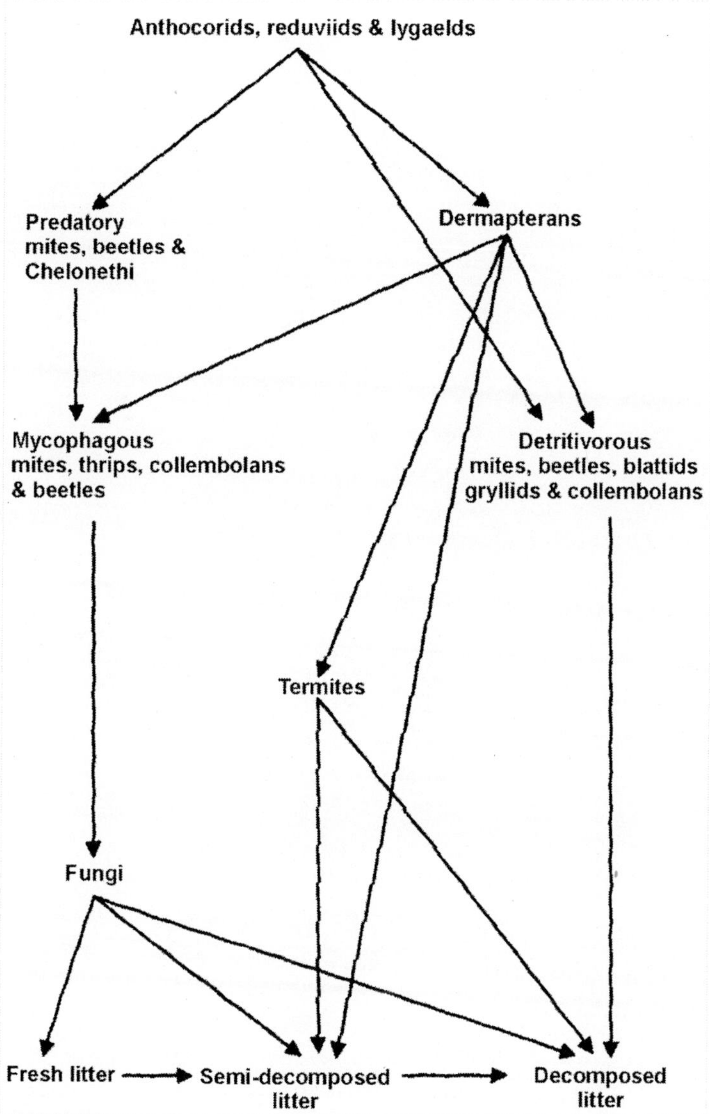

Fig. 7.6. Food relationships among major litter-inhabiting organisms

PREDATOR DYNAMICS IN THE LITTER ECOSYSTEM

The surface layer of litter abounds in macropredator species including such Hemiptera as reduviids, anthocorids, Iygaeids, Coleoptera such as carabids, staphylinids, pselaphids, earwigs or dermapterans and the Chelonethi or pseudoscorpions. Predatory collembolans, mites and thrips comprise the micropredatory elements.

Predators encounter several alternate prey in the litter ecosystems, with the advantage that they are less vulnerable to changes in the abundance of any one than a predator with a more restricted diet. The coexistence of a host of collembolans and mite predators is a fine example and from the point of view of larger predators such as reduvids, increasing tropic levels sustain their population. The F-layer in the litter system, unlike the surface layer, being comparatively stable from the viewpoint of the local environment, ensures a certain stability of the population of prey species and predators. Species may rise and fall with increments in predation intensity. All the same, the degree of diversity of prey-predator population also depends on the nature of the substrate, with changes in the relative consistency of specific composition of species.

Modes of feeding, principal food source, life cycles and habitat essentially reflect functional aspects. Taking the predators for instance, there are the general predators, specialist predators and omnivores, so that a hierarchical structure can be developed to explain interrelationships of species, as also that among feeding guilds and other functional groups.

Conclusion

Human interference with forest ecosystems in general and tropical rain forests in particular has therefore caused widespread concern regarding the consequences of such interference, which have been known to affect climate, biodiversity and human societies. While biomass is created by photosynthesis in virgin stands the process is revised during forest denudation, with the loss of litter naturally resulting in poor soil formation, reduction or loss of biodiversity, particularly the loss of ecotypes and other genetic variation. Extensive denudation in particular affects biological control resources, some of which occur at low densities, which would be vulnerable to genetic erosion or chance extinction with their reduced numbers. This would naturally affect the components of the ecosystem and the very few survivors would hardly be able to support or sustain the trophic web. The preservation and protection of forests to safeguard the natural ecosystem as well as the genetic pool they contain, are important aspects which must take into consideration the active role of litter organisms in the breakdown of litter, as well as in sustaining the food web or trophic complexities in different forest types.

CHAPTER - 8

SOCIAL FORESTRY AND INSECT BIODIVERSITY

The impact of habitat transformation on spatial abundance of insects is well reflected in the area of social forestry, an initiative by man to promote the ever increasing demand of firewood, leaf fodder, small timber and a variety of tree based produce for rural cottage industries, besides producing additional hosts for several species of major pests and thin natural enemies. Various species and types of trees have different requirements of soil and climate, such as deep or shallow, sandy and clayey, supporting different types of trees. In areas where water accumulates for longer periods only certain species can be grown some fast grown, other comparatively slow, with preference for the former. In drier parts of the country these trees which form wind-breaks and shelter belts, reduce the adverse effects of dry and desiccating hot winds which affect the yield of agricultural crops. By conserving and increasing the availability of soil moisture, these wind breaks improve crop yields. Progressive farmers in these drier belts recognize the importance of growing trees on the edges of agricultural fields and which incidentally provide firewood and fodder.

The economic importance or significance of tree planting varies from region to region depending on the utility and value of different species. Plantation of trees are being raised in the denuded lands as well as canal banks and roadside. It has been realized that the socioeconomic development of the dry region depends upon the proper management of land through large scale plantations. A large number of species has been meaningly raised, some of the most important bery *Prosopis spicigera* species of *Azadirachta indica, Pongamia glabra, Albizia lebbek, Parkinsonia aculeata.* Such tree-based economy enables coexistence with people. The scope for developing farm forestry, besides creating shelter belts, and tree avenues on road sides also enables promotion of small scale industries and cottage industries.

The need for a study of the entemofaunal resources of social forest stands cannot be overemphasized, with such studies pertaining to the nurseries as well as mature stands providing a wealth of information on taxonomic, biological and biocontrol aspects. Some of the major pests relate to *Indrabela quadrinotata* the bark eating lepidoptern borer, notably infesting *Acacia nilotica, Pongamia glabra, Albizzia lebbek; Gracillaria* sp on *Pongamia glabra, Parnara mathias* the skipper butterfly, *Eumeta crameri* the bagworms infecting Acacia, Tamarind and casuarina and *Chaliodes ferevitrea,* a pest on *Tamarind* and *Acacia* and *Pteroma plagioceps,* a serious pest of acacia, tamarinds, *Deplonix regia* and *Albizzia, Latoia lepida* the slug caterpillar causing serious damage to Tamarind and *Azadirachta indica.* Besides these are *Euproctis lunata,* the castor hairy caterpillar defoliating *Acacia nilotica, Tephrina pulinda,* the semilooper feeding on *Acacia nilotica* and *Virachola isocrater,* the fruit borer of *Tamarindus indica.* Besides, smaller species *Psylla hyalina* and *Oxyrachis tarandus,* the membracid and pod infesting coccids are common in Acacia and Tamarind.

Equally diverse are the biocontrol agents-both predators and parasites, which make the faunal diversity of social forest stands, a subject of sustained interest. To enable a proper understanding of this diversity the following tabular representation of the pest species, the host, life cycle and parasite/predators would appear useful. Surveys of a few Acacia nurseries and plantation provided information on the two species of defoliators with as many as seven parasites and six predators. This large scale biocontrol resources are due to the social forest ecosystem providing a sustained resources base for associated insects by virtue of their longer rotation periods and the pest populations and their natural enemies tend to be more persistant, entrenched in the stable habitats of the forests. Similarly bagworms form important defoliator complexes, potentially capable of causing damage to trees and acquiring pest status and 19 species are recorded from India and natural enemies play an important role in checking the build - up of bagworms. 18 species of natural enemies have been known to regulate populations of *Pteroma plagiophleps.* Among predatory species, the reduviids like *Brassivola, Eocanthecona and* coccinellids play an important role in keeping pests under check. Brief biological aspects relating to the various species of pests and their natural enemies in social forest trees are presented as below :

INDARBELA QUADRINOTATA (*LEPIDOPTERA : METARBELIDAE*)

The adult moth of this bark eating borer is whitish grey in colour, short and stout with brown streaks on the fore wings measuring about 32-35 mm. The adults are female moths vary from 7-10 and 2-6 days respectively. The eggs are laid in clusters of 15-45 under the loose bark of twigs. The incubation period ranges from 7-12 days. There are five larval instars and the full grown caterpillars are dirty and reddish brown in colour measuring about 55 mm in length. The total developmental duration (egg to adult) is about 345 days; with a larval period varied from 290-309 days, and pupal duration of about 30-36 days. The caterpillar is nocturnal, feeds on the bark under shelter of web and bores into the trunk and branches. Only one caterpillar is found inside a tunnel. Infestation is indicated by the presence of peculiar winding linnel, prepared out of excretory pellets and frass spun together with silk, generally near the forks or angles on the stem and branches.

PARNARA MATHIAS Fabr. (*LEPIDOPTERA : HESPERIDAE*)
(Skipper Butterflies)

The adult is dark brown with two white spots on the fore wings. There are five larval instars with a total duration of 16 days, followed by a pupal stage of 8 days. This butterfly passes through four to five generations per year, with a peak population during may to September. Eggs are laid singly on the upper surface of leaves. The incubation period ranged from 3 to 4 days. The newly hatched larvae first feed on the tender leaves. After 3 or 4 days the larva folds the leaves fastening the edges together with the help of silken threads produced by it, and feeds from the edges.

EUMETA CRAMERII (*LEPIDOPTERA : PSYCHIDAE*)

The bagworm *Eumeta (Clania) cramerii* is a widely distributed polyphagous defoliator of several ornamental shrubs and social forest plantations like *Acacia nilotica, Tamarindus indica* and *Casuarina equisetifolia*. The larval stages of the *E. cramerii* consists of nine instars with a total duration of 118 days. The mean pupal periopds are 21 days and 7 days for males and females respectively. They are characterized by the development of sexual dimorphism, winged male and apodous and apterous larviform females. The larvae, pupae and the adult females are found to occur inside the bags which is made up of silk thread secreated by the larvae and other extraneous materials.

PTEROMA PLAGIOPHLEPS (LEPIDOPTERA : PSYCHIDAE)

The bagworm, *Pteroma plagiophleps* is also a serious pest of *Acacia nilotica, Tamarindus indica, Delonix regia and Albizzia falcataria.* The most notable feature of the biology of *P. plagiophleps* is the extreme development of sexual dimorphism, the male being winged while the females are degenerated and larviform as in other psychids. The larval stages of both sexes inhabit portable bags constructed out of silk and plant materials. The developmental period of *P. plagiophleps* ranges from 55-75 days. The mean larval period is about 49 and 62 days for male and female respectively, while the mean pupal period is 12 and 6 days. The fecundity is 110-145 eggs/female.

LATOIA LEPIDA ***Cram.*** *(LEPIDOPTERA : LIMACODIDA)*

The slug caterpillar is a sporadic pest causing serious damage to *Tamarindus indica* and *Azadirachta indica.* The moth is stout, with green wings fringed with brown patches at the base of the each fore wings. The caterpillar is fleshy and slug-like with yellowish green body bearing a greenish blue stripe dorsally and yellowish green stripes laterally. The body is beset with series of tufts of spines highly irritating to touch, hence called `Bettle grub.' The larvae remain hidden within the leaf axis during day time and feed at night on the leaves leaving the midrib and veins. The larva becomes full grown in about six weeks undergoing seven instars, which pupates for 4 to 5 weeks, in a hard shell like cocoon on the tree trunks.

METANASTRIA HIRTA *(Cram) (LEPIDOPTERA LASIOCAMPIDAE)*

The defoliating graying hairy caterpillar, which feeds on leaves of *Tamarindus indica* and *Acacia nilotica* voraciously. The female moth has grayish brown wings and is stout. The male is smaller and has a white spot, in the centre of the black patch on the fore wing. There are five larval instars with a total duration of 40-45 days. Pupation taken place on the tree trunks inside a silken cocoon.

TEPHRINA PULINDA Walker *(LEPIDOPTERA GEOMETRIDAE)*

The semilooper moth is dirty whitish with three brownish bands, the distal one being the broadest. The larva grows to a size of about 15 mm, feed on the leaves of *Acacia nilotica,* and drops on silk threads when disturbed. They pupate in the soil or attached to the fallen leaves without forming cocoons. The larva feeds on the leaves and

Fig. 8.1. Pest damage to *Tamarindus indicus*

Fig. 8.2. Pest infestation in *Acacia nilotica*

Fig. 8.3. Larvae of *Euproctis lunata* feeding on bark of *Acacia nilotica*

Fig. 8.4. *Metanastria hyrtaca* stages

completely strips the tree when abundant. There are four larval instars, with a duration of 12-14 days and a pupal period of 4-6 days. The adult longevity is 3-4 days.

EUPROCTIS LUNATA Walk (LEPIDOPTERA : LYMANTRIDAE)

Euproctis lunata is a polyphagous pest, commonly known as the castor hairy caterpillar, also common on several social forest trees. The adult moth is pale yellow in colour and its fore wings have a black lunule. The female has a brown ochrous anal tuft. The female moth lays its eggs (95-112) in clusters on the under side of the leaves and covers them with pale brown bairs. There are six larval instars with a total duration of 26-28 days. Pupation takes place in silken web spun by the final in star larva. Pupal period lasts for 9-11 days. The total developmental period (egg to adult) takes about 50-55 days with four overlapping generations per year. *E. lunata* has been found to cause extensive defoliation of *Acacia nilotica.* Apart from *E. nilotica* a number of alternate host plants exists for *E. lunata* which includ castor, cow pea, red gram and bendhi. This insect is almost kept in check in one way or other by a number of biocontrol agents like *Apanteles cirphicoa* (Braconidae), *Eurytoma braconidis* (Eurytomidae), *Tetrastichus krishnieri* (Eulophidae), *Charops* sp, *Brachymeria* sp (Chalcididae) and a Dipteran parasite (Fig. 8.3).

VIRACHOLA ISOCRATES Fabr. (LEPIDOPTERA : LYCAENIDAE)

This is a fruit borer of *Tamarindus indica.* The adult bluish green butterfly, has an orange spot on each of the fore wings and black spots on the hind wings. The eggs are laid singly on flowers and tender fruits. The caterpillars are dirty brown and has short hairs on their body. The total life cycle duration take 40-60 days. The larva bores into the fruits, plug the bore hole with the anal segments and feeds on the internal constituents which leads to total damage of the fruit. It is found to attain high populations during the fruiting seasons in Madurai, Dindigul, Salem and Dharmapuri districts. The other food plants include guava, soapnut, woodapple etc.

SELEPA CELTIS Moore (LEPIDOPTERA : NOCTUIDAE)

The yellow hairy caterpillar *S. celtis* is a serious pest of the plantations and nurseries of *Acacia nilotica.* The moth is grayish brown with a wing span of 20 mm. The larvae is yellow with black spots and sparse, long, and pale hairs. The larvae feed on leaves of saplings in curseries and young plantations of *Acacia nilotica* and cause defoliation during rainy season. It has five to six generations per year. They are also found to attack other plants such as *Terminalia spp., Ficus carica, Solanum melongena* etc.

OXYRACHIS TARANDUS Fb (HOMOPTERA: MEMBRACIDAE)

This is commonly called as ``cow bug", which are brown and shiny, the body apparently covered with moisture; the abdomen terminates in a telescopic tube, from which issues the liquid excretions which attract the ants. The adults and the nymphs suck the sap from stem of *Albizzia lebbeck* and causes ultimate is withering and drying of the infested portions. A female lays 50-100 eggs in rows into the tissues of the tender stem and thus cause injury. It is a polyphagous pest found to have a wide range of food plants, including. Social forest trees like *Acacia nilotica, Tamarindus indica* etc.

PSYLLA HYALINA **Mathur** *(HOMOPTERA PSYLLIDAE)*

Psylla hyalina is a tiny insect (1.7-2.05 mm) commonly called as jumping plant lice, having hyaline transparent wings, long antennae and long legs. Adults are pale yellow with greenish tinge, with large eyes and the nymphs are pale green or yellow with chrome lemon tinge in the abdomen, with silvery grey eyes, creamy legs and wing pads. Both the adults and nymphs are gregarious feeders which cause curling and crinkling of tender shoots of *Albizzia lebbeck*. In severe cases they may even succumb to the attack and the saplings become stunted and dwarfish. They occur almost throughout the year with varying population sizes during different months. The eggs are laid either singly or in masses in the folds of young leaves and buds. It is also recorded on *Albizzia procera* and *Cassia siamea* in Madurai and Dindigul Districts.

Table 1. Major Pests of Social Forest Plantations and Nurseries

Species	**Plantations**	**Nurseries**
Lepidoptera		
Eumeta cramerii	*Acacia nilotica & Tamarindus indica*	*A. nilotica*
Pteroma plagiophleps	*T. indica & A. nilotica*	
Metanastria hirta	*Acacia nilotica* and *Tamarindus*	
Manatha assamica	*T. indica & A. nilotica*	
Chaliodes ferevitrea	*T. indica & A. nilotica*	*A. nilotica & T. indica*

Species	Plantations	Nurseries
Euproctis lunata	*A. nilotica*	*A. nilotica*
Tephrina pulinda	*A. nilotica*	*A. nilotica*
Parnara mathias	*P. glabra*	*P. glabra*
Gracillaria sp.	*P. glabra*	-
Latoia lepida	*Tamarindus indica* *Azadirachta indica*	
Virachola isocrates	*T. indica*	-
Indarbela quadrinotata	*P. glabra* *A. nilotica & Albizzia lebbeck*	-
Selepa celtis	Acacia nilotica	
Coleoptera		
Coelosterna sp.	*A. nilotica*	-
Myllocerus viridanus	*A. nilotica & T. indica*	*A. nilotica & T. indica*
Myllocerus curvicornis	*T. indica A. nilotica & P. glabra*	*A. nilotica & T. indica*
Myllocerus tenuicornis	*Azadirachta indica*	*A. indica & T. indica*
Orthoptera		
Orthacris maindroni	-	All the nurseries
Atractomorpha crenulata	-	All the nurseries
Homoptera		
Helopeltis antonii	*Azadirachta indica*	-
Psylla hyaline	-	*A. lebbeck*
Coccids	-	All the nurseries
Larval Parasites	**Host insects**	
Sinophorus psycheae	*Eumeta cramerii , Chaliodes ferevitrea & Manatha assamica*	
Tachinid	*Eumeta cramerii, Manatha assamica, Chaliodes ferevitrea*	
Phanerotoma sp.	*Thyridopteryx Ephemeraeformis*	
Enicospilus sp.	*Tephrina pulinda* *Euproctis lunata*	

Species	Plantations	Nurseries
Apanteles cirphicola	*Tephrina pulinda* *Euproctis lunata*	
Acanthormius sp.	*Pteroma plagiophleps*	
Charops sp.	*Euproctis lumata*	
Tetrastichus krishnaieri	*Euproctis lunata*	
Eurytoma braconidis	*Euproctis lunata*	
Pseudagenia sp.	*Metanastria hyrtaca*	
Pupal Parasites		
Goryphys sp.	*Pteroma plagiophleps*	
Brachymeria carinata indica	*Pteroma plagiophleps*	*Tamarindus indica*
	Manatha assamica & Eumeta cramerii	*Acacia nilotica*
Brachymeria mormonti indica	*Pteroma plagiophleps*	*Tamarindus*
	Manatha assamica & Eumeta cramerii	*Acacia nilotica*
Brachymeria criculae indica	*Metanastyria hyrtaca*	*Tamarindus & Acacia nilotica*
Kriechbaumerella ayyari indica	*Latoia lepida*	*Tamarindus*
Predators		
Chrysopa scelestes	*Coccids & eggs of Parnara mathias*	*Pongamia glabra & Azadirachta indica*
Coccinella transversalis	*Coccids, Aphids & Psylla hyaline*	*Acacia nilotica Albizzia lebbeck & Azadirachta indica*
Eocanthecona furcellata	*Latoia lepida*	*Tamarindus indica*
Rhinocoris fuscipes	*Tephrina pulinda & Parnara mathias*	*Acacia nilotica & Pongamia glabra*
Irantha armipes	*Tephrina pulinda*	*Acacia nilotica*
Brassilova sp.	*Euproctis lunata & Eumeta cramerii*	*Acacia nilotica*
Oxyopes rufisternum nilotica	*Tephrina pulinda*	*Acacia*
Peucetia viridana	*Tephrina pulinda*	*Acacia nilotica*

CHAPTER - 9

INSECT DIVERSITY IN MOSSES, LICHENS AND FERNS

Introduction

Insect life in mosses, lichens and ferns has in recent years attracted the attention of entomologists in view of their richness and adaptive diversity in various environments. Bryophytes have diverse life forms and grow in different colonies in a variety of habitats, often forming cushions, with diverse groups of insects associated with them. Lichens are also important components of natural habitats from rocky substrata to tree barks, not to mention of the abundance of fern species and the insects that colonize them. Information relating to insect interactions on these hosts from India is comparatively meagre, calling for a concentrated effort to investigate insect-lichen-moss and fern interactions, not only from the viewpoint of their diversity, but also pertaining to some aspects of their chemical inputs in relation to their interactions. The diversity of insect life in these environments has in recent years attracted quite some interest from the viewpoint of adaptive diversity (Gonsalez *et al.,* 1998).

Insect Diversity on Ferns

It is generally assumed that ferns are not eaten by insects, Ehrlich and Raven (1964) remark that ``Very few insects, indeed, feed on ferns at all, a most surprising and as yet unexplained fact". In today's perspective many of the ferns show a considerable amount of damage due to arthropod feeding and hence the widely held assumption appears not well founded. As early as 1922, Swezey studied insects on ferns in Hawaii. In the manual on Pteridology, Leeuwen (1938) provides an extensive list of fern insect pests. One of the most exhaustive accounts on fern insects is provided by Balick (1978), where on the basis of literature, he cites 420 named species of arthropods, besides an account of the biochemical and evolutionary aspects of arthropod predators on ferns, certainly a work of reference. More insects are present on ferns than earlier envisaged and many have overcome their chemical defense system to utilize them as food. While Hendrix (1980) notes that greater number of ferns are utilized by insects in the Hawaian islands, the recent work of Lawton (2000) on *Pteridium aquilinum,* an economically important bracken, assumes importance, as the best known pteridophyte, wherein he identified forty species of regularly or occasionally feeding insects, of which twenty are core species of regularly or occasionaly feeding insects, nine strictly monophagous, nine oligophagous (both bracken and other plants) and nine polyphagous, feeding on ferns and other diverse plants.

Concentrated efforts at studying insect ferns in India have been negligible, the only available literature pertaining to it being by Ghosh (1974) on fern infesting aphids. Priesner and Seshadri (1952) recorded *Indusiothrips seshadrii,* while Ananthakrishnan (1968) reported *Scirtothrips pteridicola,* besides several species of *Xylaplothrips* and *Stigmothrips* on dead ferns. *Heliothrips haemorrhoidalis* was recorded on the fern, *Polypodium phaegopteris,* restricted to mature fronds where they feed on the abaxial and adaxial surfaces, leaving white silvery pathces with numerous blackish markings caused by the deposition of faecal matter by the larvae (Ananthakrishnan and Gopichandran, 1993). Similarly, scattered information on mites and whiteflies are available. Many have overcome the ferns' chemical defense systems to utilize them as food and needless to say that insect fern interactions provide a good example of coevolution.

Approximately 420 named species of arthropods have been recorded as feeding on 44 fern species, the majority of which are from the orders Coleoptera, Hymenoptera, Lepidoptera, Hemiptera, Homoptera, and Thysanoptera. Evolutionarily primitive sawflies and advanced moths are reported to be present in ferns, suggesting possible co evolution of ferns and arthropods both before and after the radiation of angiosperms (Balick *et al.,* 1978). Tannins and cyanogenic glycosides determine the extent of insect attack on a fern (Cooper Driver *et al.,* 1977). The types of damage observed relate to small crescent of notching of pinnules by curculionids, large notching of pinnae by Lepidoptera, rarely Orthoptera; skeletonization of upper surface of pinnules between veins by Lepidoptera and Coleoptera; a short hole on the upper surface by Hemiptera and Homoptera;

complete or partial elimination of several pinnules on one or both sides or rachis by Lepidoptera. As such, the modes of feeding include - sporangia feeders, gall formers, rhizome feeders, miners and borers and decaying frond feeders. The largest group of feeders are Coleoptera, Hymenoptera and Lepidoptera. Weevils are widely associated with ferns all around the world especially in the tropics. Among Lepidoptera, Noctuidae are the most numerous fern feeders containing many specialist fern feeders. Aphids are able to utilize a large number of divergent species and Eastop (1973) records 29 species of aphids on ferns. Diptera and Thysanoptera are restricted fern feeders. Aquatic ferns like *Salvinia* are known to be attacked by *Bagous* and *Cyrtobagous.* An interesting fern from Florida, is *Lygodium microphyllum* a climbing fern, makes its way up the sterms and trunks of trees. On the ground, it creates tough, spongy mats, infesting over thousands of acres. The moth, *Cataclysta camptozonale,* an Australian native is a biocontrol agent, preferring to eat the fronds. Active research on this blanket fern has led to the discovery of several biocontrol agents, Climbing fern is also a native of South East Asia viz., Indonesia, Malaysia and Thailand. A mite, *Floricanus* sp. feeds on the leaves, causing leaf fall, besides a leaf mining beetles (USDA - ARS - 2002). The fern scale, *Pinnaspis aspidistrae* infesting the upper and lower surface and on stems, is a cosmopolitan species :-

An understanding of the distribution patterns of fern species at different elevations and a comparison between local guilds and assemblages across sites appear essential elements. However, of the several species of ferns known from India, some of the more common are: *Psilotum, Lycopodium, Isoetes, Lygodium, Pteris, Adiantum, Pteridium, Nephrolepis, Asplenium, Marseilia, Azolla and Salvinia.* The insect fauna on these ferns needs to be investigated so as to have an overall idea of the fern entomofauna. The only exhaustive list is aphids from Indian ferns by Ghosh (1974) (Table 1). Plant assemblages always tend to vary in their diversity, their composition and biomass variation being determined by climatic variation. There is a need to study the ecology of fern feeding insects as done by Lawton (2000). He has shown that species richness, dominance and spatial arrangement, the length of food chain and organization of food webs are importance criteria, being very typical of community ecology. It would perhaps be relevant to cite some common species of insects such as aphids, thrips, Lepidoptera and Coleoptera known from Indian ferns, to enable further incisive studies.

Some common fern infesting insects

S.No.	Name of the arthropod	Family	Order	Name of the Fern
1	*Cataclysta camptozonale*	Pyralidae	Lepidoptera	*Lygodium microphyllum*
2	*Neomusotima conspurcatlis*	Pyralidae	Lepidoptera	*Lygodium microphyllum*
3	*Musotima* sp.	Pyralidae	Lepidoptera	*Lygodium microphyllum*
4	*Callopistria duplicans*	Pyraustidae	Lepidoptera	*Lygodium microphyllum*
5	*Neostromboceros albicomus*	Tenthridinidae	Hymenoptera	*L. microphyllum, L. flexuosum*
6	*Amphorophora ampullata bengalensis*	Aphididae	Hemiptera	*Cheilanthes* sp., *Pteris aquilina*
7	*Anthracosiphoniella maculatum*	Aphididae	Hemiptera	*Asplenium esculentum; Athyrium*
8	*Bagous* sp.	Curculionidae	Coleoptera	*Salvinia* sp.
9	*Macromyzus polypodicola*	Aphididae	Hemiptera	*Dryopteris arida; Polystichum* sp.
10	*Floracarus perrepae*	Eriophyidae	Acarina	*Lygodium microphyllum*

According to Hendrix (1980), host switching accounts for the diversity of insect feeding on ferns, in particular the root, stem and foliage feeders, Balick *et al.* (1978) have earlier reported root feeders (3.3%), spore feeders (2.4%), miners or borers (5.9%), gall forms (4.3%) and decay feeders (4%). In their view, 73% are restricted to ferns and 27% are polyphagous. These are aspects which need incisive studies in respect of Indian fern insect fauna, with regard to varying tannin levels, and with more rapid coevolution of both insect and fern hosts, the insects tend to develop a high degree of tolerance to tannins and this perhaps a reason as to why ferns with high levels of tannins are still subject to insect attack (Balick *et al.,* 1978). Ferns like *Polypodium* and *Pteridium* possess a wider range of deterrent toxic compounds such as cyanogenic glycosides and sesquiterpenoid lactones and hence have fewer members of insects. The need for a correlation of fern insects in terms of fern sporophagy involving switching from mycophagy to sporophagy, though debatable, needs

Fig. 9.1. *Lygodium microphyllum*
Left: Climbing a coconut tree; Right: Close up of *Lygodium*
(After Alexander Jesudasan)

Fig. 9.2. Insect feeding on fern (After Alexander Jesudasan)

Fig. 9.3. Insect feeding on fern (After Alexander Jesudasan)

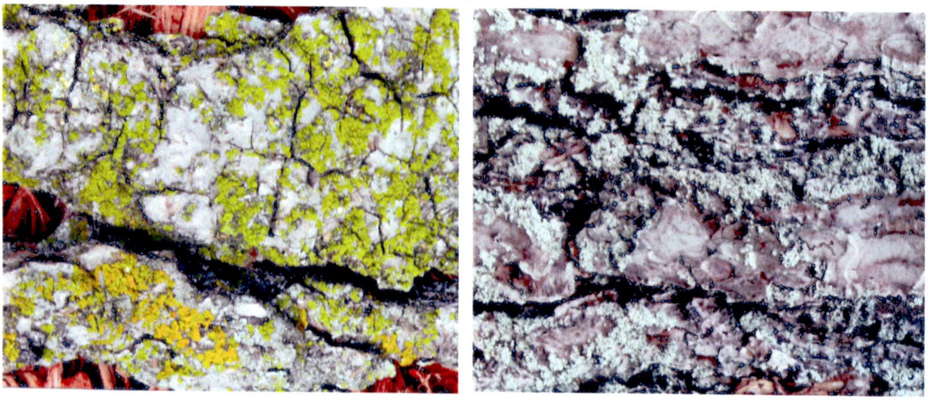

Fig. 9.4. Moss encrustation on bark

Fig. 9.5. A general view of moss encrustation

Fig. 9.6. Lichens on bark (After G.N. Hariharan)

Fig. 9.7. Foliose lichens (After G.N. Hariharan)

more critical studies on the special adaptation, as perhaps occurring in Coleoptera (Leschen and Lawrence, 1991).

Moss insects

Bryophytes or mosses have been called `the amphibians among plants' by Richards (1954) as they straddle the aquatic and terrestrial worlds. They are therefore stepping stones or half way houses for several invertebrates. The ability of bryophytes to live under ``intermittent wetting and drying'' is presumably significantly in this regard. Mosses are among the most conspicuous among the early soil colonizers and important components of high mountain ecosystems (Mani, 1962). Insect abundance is related to the shape, texture and thickening categories : *bryobionts,* occupying exclusively in association with bryophytes; *bryophiles,* usually found on bryophytes, but may survive elsewhere and *bryoxenes,* which spend part of their life cucle on bryophytes (Gerson, 1982; Smith, 1982). As different bryophytes have different life forms and grow in different colonies, the shape, size and texture of leaves affect the associated fauna.

Urban areas are characterized by high levels of disturbance and environmental modifications so that knowledge of vegetation dynamics is basic to our understanding of ecological interaction. Besides increasing loss of herbivorous communities and replacement initially by bare earth, followed by grasses, reduction in the biomass of trees and shrubs and the loss of biodiversity through severe structural modifications coupled with fragmentation, results in course of time, the vegetation being no longer related to environmental vagauries. As vegetation changes with succession, so does phytophages increase with their niche breadth declining as succession proceeds and in course of time, generalists tend to replace specialists. The broad pattern of arthropod invasion is as follows : ants, butterflies, carabids, Hemiptera, Orthoptera, bees.

Insect-Moss Associations

Under favourable conditions, bryophytes form perennial colonies in many environmental, thereby becoming apparent as predictable environment to prospective feeders. Changes in bryophyte composition affect the distribution of insects. The need for a better understanding of insect moss association in the Indian context is called for, besides the impact of bryophyte secondary substances such as terpenoids, pineone, limonene, sitosterol and stigmasterol. As such the brief list of insects presented here are essentially taken from Gerson (1982) so as to provide a base for future studies on moss entomology in India. Most cushions developing on rock surfaces were first colonized by rhizopods, rotifers and tardigarades and as dead material formed under the cushion, insects like collembola and oribatid mites increased. Diverse insects feed on bryophytes, with Orthoptera (grasshoppers), beetles and caterpillars biting and chewing and bugs, aphids, and mites sucking cell contents. Collembola like *Sminthuroides, Folsomia* and *Isotoma* are common besides swarms of thysanopterans feeding on dry mosses and lichens at high altitudes (Mani, 1962), besides proturans. Moss springtails occur in large numbers, the estimated numbers being over 2000000/m^2 in mosses. Species of *Xenylla* appear to be bryophilous. Aquatic mosses abound in *Ephemerella* and Baetis and dragonflies like *Calicnemian* are Himalayan species feeding on aquatic bryophytes (Kumar and Prasad, 1977). Stonefly populations reach substantial numbers like *Nemurella* (Lindeyard *et al.,* 1975). Crickets and grasshoppers feed on *Sphagnum* and some Orthoptera feed on thick growths. Rare instances of Embioptera and some stick insects are known to feed on mosses.

Moss feeding thrips is not uncommon and Mound's (1989) review indicates that they are restricted to a single tribe, *Williamsiellina* with 25 species, common in the New World, although widely distributed, besides 17 species of *Lissothrips,* the common species being *Lissothrips muscorum.* They feed on leaf and flower tissue of higher plants including pollen grains as well as fungal hypae and spore, besides mosses. The tubuliferan genera, *Lissothrips, Pththirothrips, Microlimothrips, Prolimothrips* and Trisclerothrips are known from mosses and among the Terebrantia are *Bournierothrips* and *Wegenerithrips. Lissothrips okajimi* was collected from moss on a live tree in Japan.

The hemipteran family, Peloridridae feed on mosses, besides several aphids. Leston (1959) listed bryophyte associated bugs, which attach their eggs to moss stalks. Lacebugs of the genus, *Acalyphus* feed on bryophytes. Of the beetles, staphylinids occurs in mosses. In some beetles the dorsum becomes extensively overgrown by bryophytes such as liverworts and mosses. Chironomids abound in submerged mosses, besides tipulid flies. Caddisflies like *Phyacophila verrula* live in bryophytes and construct cases in liverworts. Diptera are the most closely associated with bryophytes. Ants like *Myrmica* and *Formica* form nests in sphagnum. Moss mites such as oribatids are very abundant, and common on Sphagnum mats. Insects therefore have been known to play a role in bryophyte evolution. Many insects camouflage themselves with the mosses such as caddisflies, their cases composed of leafy liverworts.

In short, a variety of flies, rich in diversity and abundance occur in mosses, being adapted to lanchaceous mosses in particular. Diptera are consistently associated with mosses, since no other order of insects present se great a diversity of larval habitats as Diptera (Richards and Davies, 1977), bryophytes meeting the needs of Diptera in ways more than one. Sawflies, mosquitoes, simulids and certatopogonids frequent mosses, besides coprophagous and dung inhabiting flies, which are attracted to the capsules of certain Splanchnaceae. The presence of specific alluring chemicals, the distribution and growth sites of mosses and relatively large numbers of entomophilous flies attest to the diversification and evolution of splanchnaceous mosses (Gerson, 1982). Insects living in these environments have evolved means to overcome the defensive chemicals such as oxalic acid, lauric acid, limonene and lactones to mention a few which tend to repel caterpillars and mites. As there is no specialization in terms of moss hosts, many insects tend to feed on several, unrelated hosts and in this process are aided by mixed function oxidases, Pseudoscorpions like *Microbisium* and *Neobisium* are associated with mosses. Ants gnaw moss capsules and form nests on Sphagnum. Diverse insects feed on bryophytes - Orthoptera, beetles, caterpillars which bite and chew, bugs aphids and mites which such cell contents. Larvae of some flies such as *Dolichopeza* tunnel in bryophyte rhizoids and those of *Lycosia* in the thallus of *Marchantia* (Byers, 1961). The effect of bryophytes on stream invertebrates is of interest in that mosses increase the densities of *Ephemerella* and chironomids. In complex aquatic microhabitats involving the green layers, nymphs of mayflies like *Baetis* and *Ephemerella* occur.

Lichenivorous Insects

Lichens are important components of natural habitats often forming encrusting carpets on the groud or trees, providing nutrition and acting as good indicators of environmental health. Lichens are symbiotic organisms of fungi (mycobionts) and algae (Photobionts), commonly grow on rock surfaces, poorly developed soils and as epiphytes on trees and shrubs, providing an important potential source of food for many insects. The general resistance of lichens to insects attack, is due to the presence of lichen phenolics. An interesting example is the association of a butterfly. *Talicada nyseus* and the lichen, *Leprolema sipimanianum*, the lichen being restricted to montane habitats and rare in tropical rainforests, the larvae feeding on the lichen. Many tree trunks are mottled with green and greyish growth of lichens and mosses, and bark living bugs and mantids blend imperceptibly with the background. Lichenivores are relatively common in nature, ranging from small invertebrates such as spring tails or Collembola, mites and lepidopteran larvae. Paucity of work relating to Indian species of insects has led to insufficient information relating to lichen-insect interactions. As such exotic examples are provided. Lichen moths belonging to the family Arctiidae occur allover the world and over 200 species of lichen moths are known from Australia. Some of the common species are *Termessa gratiosa,* the banded lichen moth. *Eutane terminalis,* the lichen eating caterpillar, *Manulea replana,* the lichen moth. *Scoliocma nana* and *Termessa catocalina* being common in Australia. It is interesting that lichen moths are good indicators of environmental pollution.

Other lichenivorous species feeding on epiphytic lichen species are four species of *Eilema* on the lichen, *Hypogymnia physodes, Melanetia exaspirata, Vulpicida pinnastris, Xanthoria parientina.,* known from Finland. Lichenivory in Lepidoptera therefore occurs in several families, the largest radiation of lepidopteran lichenvores occurring in the family Arctiidae (Raulins, 1984).

Lichens produce a variety of lichen products having a wide range of biological activities. Moths of Arctiidae use lichen phenolics for defense purposes. In lichens, secondary chemicals are of fungal origin, and are mainly deposited only on the surface of the hyphae. Since many lichens contain relatively high concentrations of secondary chemicals, arctiid moths feeding on them, have a relatively slow development. Lichens show a wide range of sensitivity to So_2 and *Lecanora conizoides* is the most toxic tolerant epiphytic lichen known.

Some fruiticose lichens found on exposed tree trunks display variability in growth form, influencing the habitat of corticolous arthropods such as Pscoptera, species of *Entomobrya,* small spiders, the thalli acting not only as a fond resource, but also helping to retain humidity to provide shelter. Grazing by Psocoptera and oribatid mites and the entomobryid collembolan, *Orchesella* sp. also induce variability in the growth of lichens.

CHAPTER 10

INSECTS AND THE GALL MICROENVIRONMENT

Galls are typical growths induced by a wide variety of organisms from viruses, bacteria to insects. They are highly host and organ specific, induced abnormal outgrowths in various plants parts, roots, stems, leaves or buds, but occurring mostly on leaves. Galls are induced by diverse groups of insects such as thrips, psyllids, aphids, coccids, weevils, midges, fruitflies, agromyzids, chalcids and cynipids, to mention the more important. Most galls are host or organ specific, inducing galls only on one species or a closely related group of host species. The form of the gall varies, from slight depressions or pits to elaborate growth. Irregular growths are often termed *kataplasma* galls, but more complex one are the *prosoplasma galls* their tissues undergoing differentiation as the gall matures. The basic gall types include the pit or blister galls, having undergrowth of filz galls, pouch galls, roll and fold galls and bud and rosette galls. The basic stages of gall development are in vitation, growth and differentiation, maturation and dehiscence. Gall growth and development result in increased biomass of the gall by cell divition or *hyperplasy* and cell enlargement or *hypertrophy*. Gall forming insects have developed several strategies for successful survival and both hypertrophy and hyperplasy are adaptive phenomena resulting in the establishment of gall forest. Transformation of differentiated tissues into mcristematic tissues is also an adaptive trend. Gall formation is also considered an adaptation for the exploitation of nutritionally inferior plant tissue into an organized nutritive zone, either in the form of isolated cells or a highly organized nutritive zone surrounding the insect, so that gall insect-plant association is an advanced or specialized trophic strategy. The unusal concentration of highly nutritive substances within the gall tissues and the presence of meristematic tissues provide a favourable habitat for the development microcommunities from bacteria and fungi to insects and mites. Tissues closer to the gall formers show cytological or morphological changes which are beneficial to the feeding and development of gall insects. For instance, the nutritive cells in young galls have denser cytoplasm, increased vacuoles often fragmented into smaller vacuoles, larger and modified chloroplasts. As such, galls are known to act as physiological sinks for drawing resources from the surrounding tissues. The nutritive cells present varying profiles even within a single gall system, so that the layer of cells lying close to the feeding insect shows profound structural specialization and functional adaptation. On the other hand cells away from the insect show more or less normal profiles with minimum structural alteration.

Fig. 10.1. Some thrips galls

Insect diversity in plant galls : Thrips (Figs. 10.1, 10.2)

Gall forming trait in Thrips is dominant among species of the family Phlaeothripidae with about 300 species inducing galls on diverse host plants. About 75% of the gall inducing thrips belong to the genera *Liothrips, Gynaikothrips, Liophlaeothrips, Eugynothrips and Mesothrips.* The most commonly galled hosts belong to the families Araliaceae, Melastomacea, Myrtaceae, Piperaceae and Vitaceae (Fig. 10.1). The types of galls induced by various species of thrips are :

(a) Leaf folds as in *Ficus retusa* caused by *Arrhenothrips ramakrishnae*

(b) Marginal partial rolls by *Aneurothrips punctipennis or Cordia* sp.; *Alocothrips hadrocerus* on *Maytenus senegalensis.*

(c) Marginal complete rolls caused by *Liothrips karnyi* on *Piper nigrum; Lygothrips jambuvasi* on *Anogeissus latifolia.*

(d) Laminar complete rolls caused by *Gynaikothrips, flaviantennatus* on *Casearia tomentosa Eothrips coimbatorensis* on *Jasminum pubeseens* and laminar complex galls caused by species of *Crotonothrips.*

(e) Pouch galls caused by *Austrothrips cochinchinensis* on *Calycopteris floribundus* and *Dixothrips onerosus* on *Terminalia chebula.*

(f) Horn galls caused by *Liothrips ramakrishnae* on *Schefflera racemosa.*

(g) Rosette galls caused by *Thilakothrips babuli* on *Acacia leucophloea*

(h) Tubular galls on *Lionicera* caused by *Mesicothrips plicans*

(i) Stem galls on *Casuarinia stricta* caused by *Thaumatothrips froggatti*

(j) Simple leaf curl galls caused by *Cercothrips nigrodentatatus* on *Planchonia valida.*

The size and structure of many thrips galls are dependent on the number of thrips present. Galls induced on young, differentiating leaves become complete galls, whereas those on mature leaves develop into partial galls. Large galls such as the pouch galls of *Calycopteris floribundus* often contain upto 5000 individuals and in a rare case of a species *Tolmetothrips granti* from the Solomon islands, is known to harbour more than 10000 individuals. Galls that develop from the activities of more than one species of thrips are called company galls.

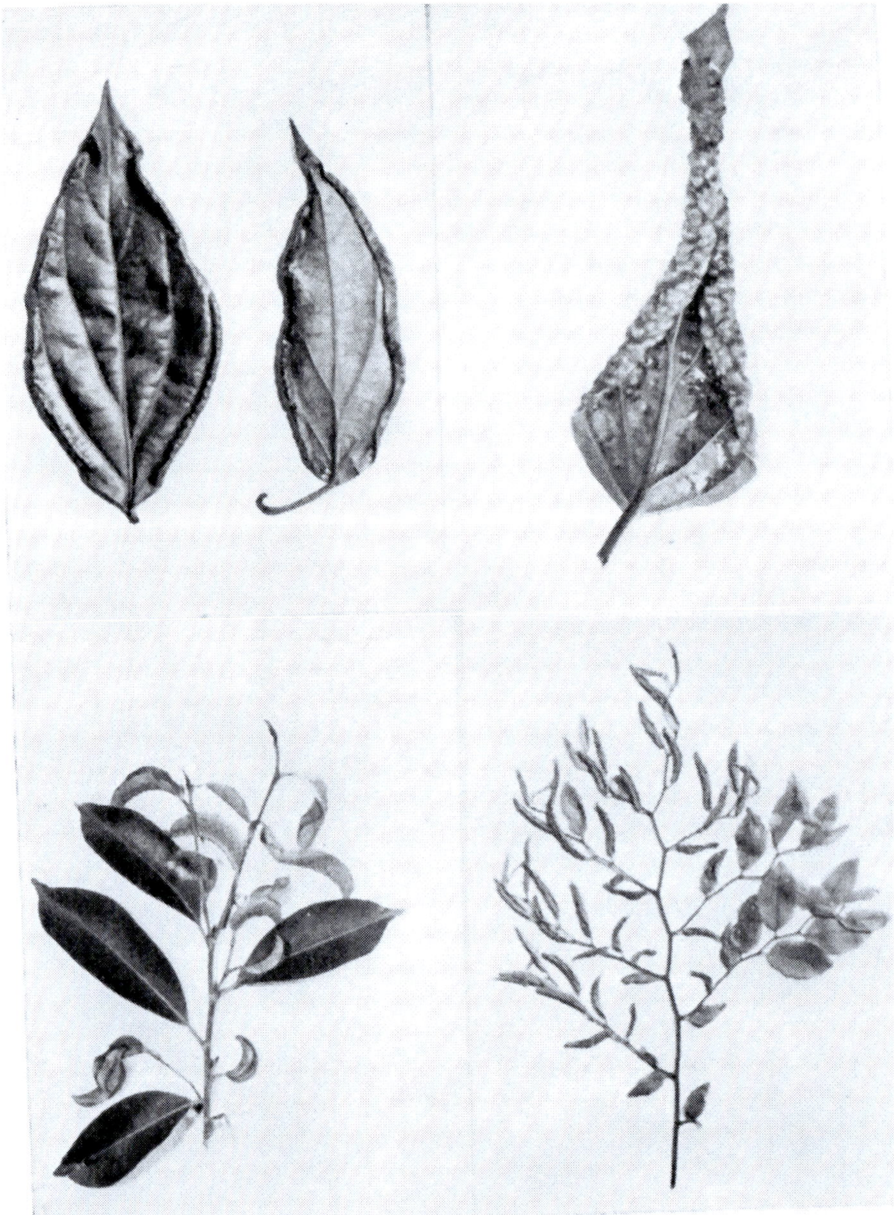

Fig. 10.2. Some thrips galls

Cecidomyiid galls (Figs. 10.3, 10.4, 10.5)

Adaptive strategies of Cecidomyiids (Diptera) involve diverse sites for gall production such as shoot, leaf, bud, stem and fruit. One of the common galls is the *silvershoot* gall of the paddy gall fly, *Orseolia oryzae*. Galls very similar to these also occur on the grass *Cynodon dactylon* caused by *Orseolia cynodontis*. These are composite galls consisting of 5-30 units of tubular galls on 10-20 leaves grouped together at the extremities of shoots. A similar gall is formed in *Paspalum by Orseolia paspali.* Among the several galls of cecidomycids, those initiated on the leaves of mango, *Mangifera indica* are of interest. Eleven species are known to induce different types of gall. Echinate galls on mango leaf are formed by *Amaradiplosis echinogalliperda.* Pustulate galls are formed by *Procontarinia matteina* and *P. mangiferae.* Besides there are species of *Erosomyia mangiferae and E. indica.*

Fig. 10.3 Cecidomyiid galls

Galls on the leaflets of *Acacia* are of interest with around 22 species known as in *Acacia suma* and *A. catechu* and *A. leucophloea.* *Schizomyia orientalis* form galls on *A. leucophloea,* while *Contarinia bivalviae* form barrel-rhaped galls on *Acacia catechu,* while *Contarinia ramachandrani* form *Cylinder-piston* galls on *Acacia suma.* The galls of *Horidiplosis mathuri* occur on the leaves of *Ficus glomerata,* while *Phyllodiplosis jijubae* forms peculiar and unusual leaf rolls on *Zizyphus jujuba.* Ficus religiosa is known to harbour leaf galls formed by *Pipaldiplosis papaldiplosis,* while *Asphondylia Phyllanthi* causes galls on *Emblica officinalis.*

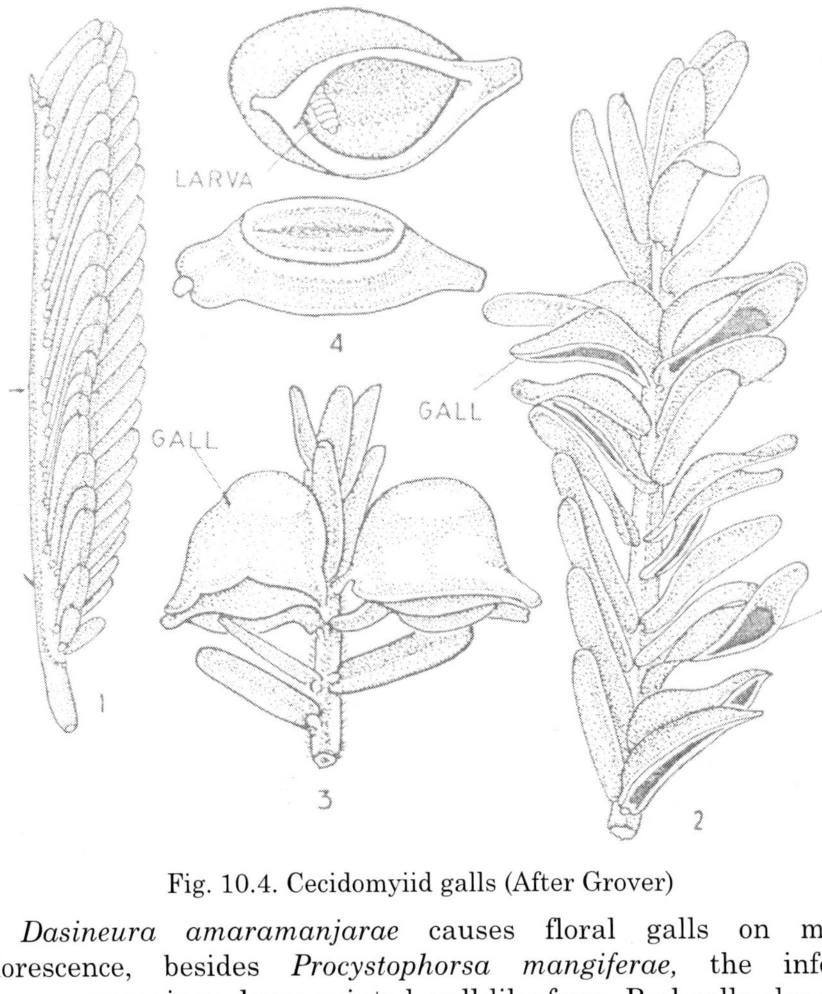

Fig. 10.4. Cecidomyiid galls (After Grover)

Dasineura amaramanjarae causes floral galls on mango inflorescence, besides *Procystophorsa mangiferae,* the infested blossoms assuming a large pointed, gall-like form. Bud galls also occur on several species, some of which are *Contarinia sorghicola* in Sorghum, *Dasyneuracitri* on citrus inflorence, *Ametrodiplosis pareirae* on *Luffa cylindrica* to mention some of the common ones.

Stem galls occur on *Coccinia indica* and *Luffa acutangula,* caused by species of *Lasioptera* are thick, tasiform, elongate, curved, hairy, fleshy galls.

Adult midges of *Erosomyia indica* cause fruit galls, while *Udumbaria nainiensis* cause fruit galls on *Ficus carica.* Round galls of *Cyamopsis tetragonoloba* are caused by *Asphondylia cyamopsi.* Among the Homoptera, the aphids, psyllids and coccids are known to be gall formers. The gall forming aphids are confined to the family *Eriosomatidae, Adelgidae and Phylloxeridae.* Many aphid species cause leaf or stem deforestation and are often pseudogalls, since they are not enclosed.

Fig. 10.5. Galls formed by *Orseolia oryzae*

Psyllid galls

Psyllids are well known gall formers, the common species being *Trioza jambolanae* on the leaves of *Syzigium cuminum, luids Trioza fletcheri* on *Terminalia arjuna* form sub galls; while *Apsylla cistellata* form bud galls on apical shoots of mango. *Trioza pitiformis* causes pit galls on *Shorea robusta,* while *Megatrioza pallida* forms leaf, flower or fruit galls on *Mallotus phillipinensis. Cecidopsylla schimae* indices marginal leaf rolls on *Schima wallichii,* while *Phacopteron lentiginosum* causes galls in the terminal leaves of *Garruga pinnata. Pauropsylla depressa* is known to form galls on *Ficus glomerata.*

Coccid galls

Many coccids induce simple galls such as bud deformation, leaf wrinkles and pit galls, through complex galls are also known, such as on Myrtaceae and Casuarinae. Species of *Gallococcus* induces epiphyllous bud galls on the leaves of *Shorea.* Males and females of coccids such as *Apirpha* induce separate, dissimilar galls on leaves of *Eucalyptus.* The males are usually alate, while the females are apterous. *Beesonia dipterocarpi* cause irregular shaped woody galls.

Gall formation

In almost all instances of cecidogenous insects, galls are induced on juvenile plant organs, though rarely on mature stems or leaves. From the nature and the morphology of the galls formed, it is evident that gall-insects are clearly species-specific. The mode of gall induction differs significantly among the groups of gall formers, the nature of their mouth parts often playing an important role.

Thrips galls develop as a result of the cumulative feeding effect of a population, although during the initial stages of cecidogenesis, very few individuals occur. Population studies on gall thrips (Raman *et al.*, 1978; Varadarasan & Ananthakrishnan 1981; Ananthakrishnan 1983) have shown that thrips abundance as well as their population build up are distinctly related to the availability of primordial plant tissues (leaf tissues, mostly). Despite their occurrence throughout the year, many gall thrips display population peaks only when the hosts produce new shoots. Feeding stimulus of thrips induces galls, and initially a few epidermal and subjacent cells are disrupted and they respond to the stimulus, that the cells begin to proliferate and this contribute to cecidogenesis. Depending upon the nature of the stimulus as also the time and location of infestation, the proliferated tissue grows into specific gall types, either as epiphyllous or hypophyllous rolls or folds, or as ceratoneon or cephaloneon pouches.

With increase in the population of concerned thrips species, the galls many subsequently attain increasing complexity.

Galls induced by aphids also synchronise their developmental stages with the availability of new host flush with tender buds, and more importantly with the season. The latter appears significant as most of the cecidogenous aphids exist in temperate regions of the world, that show clear-cut seasonal patterns. The nature of development and infestation of *Adelges* attacking the shoots of *Picea*; reaching the host tissue in autumn and overwintering is an important prerequisite for the galls to be formed in the succeeding spring (Rohfritsch, 1981). The `stinging' effect of the feeding nymphs initiate galling. The adelgid nymphs feeding on the tender shoots of the susceptible conifer, cause the swelling and coalescence of the stem region and also the needle bases in close vicinity with separate chambers for enclosing the gall aphids.

The impact of seasonal and other environmental factors does not become manifest among the galls of psyllids and coccids, probably in view of their distribution being predominantly confined to the topical regions of the world. Solitary nymphs of one gall coccid or gall psyllid feed initially on the superficial layers of the host tissue that induce the cecidogenetic responses.

The larvae emerging from the eggs laid by the female gall-midge (Cecidomyiidae) would gnaw their way into such tissues of the host organ as the stem leaf or midrib or petiole or sometimes flowers, taking a tangentially curved path. Their feeding activity, soon after settling initiates gall development characterized by the organization of a distinct nutritive tissue, and a circumscribing sclerotic zone whose function is debated. While in most of the Lasiopterini, the larvae turns into the adult within the gall, in the rest of the cecidomyiids, the gall, drops down to the soil facilitating larval diapause, wherein the life cycle is completed.

Differing significantly from the rest of the gall-makers, the female cynipid (gall-wasp) deposits the eggs within the host tissue using its long ovipositor. The larvae emerging from these eggs initiate gall developement, although the irritation caused by the ovipositor and the associated secretions were earlier considered to be responsible for gall induction. The feeding impact of the inhabiting larvae triggers gall growth into a one-chambered (unilocular) or into a multi-chambered (plurilocular) gall, with their characteristic types of nutritive tissues. Some of the cynipine galls remain persistant on the host organ till the end, while a few others drop down to the soil.

CHAPTER - 11

AQUATIC INSECTS IN ODD ENVIRONMENTS: FROM SNOW MELT WATER STREAMS TO TEMPORARY PONDS

Introduction

Entomologists all over have long been enthralled by the diversity of morphological, physilogical and behavioural adaptations of insects living in or on freshwater for part or whole of their life span. They have contributed substantially to promote the biotic integrity of freshwater ecosystems by evolving unique suvival strategies to explore and exploit an array of habitats-both macro-and micro, available from the largest lakes and rivers to smallest ponds, ditches, springs, streams, coconut shells, tree holes, discarded tyres and cans and even animal footprints. Their species richness is directly proportional to the habitat heterogeneity, as for instance, in the stream environment, they are often associated with the channel botton or with such stable surfaces as fallen leaves and submerged or emergent aquatic vegetation. Temporary waters in general and ponds in particular are noted for their wide variety of speceis, forming part of the functional communities, their ecology being of interest in view of their being smaller habitats. Intermittent ponds and ditches and irrigation canals support large numbers of mosquito larvae. Small water bodies are often classifiable into puddles, pools and small ponds. Three types of life cycles occur in temporary waters viz., (a) rapid growth occurs at the beginning of the pool stage; (b) an even growth rate spreading over the entire pool stage with such insects as *Gerris, Anisops,* dragonfly nymphs and *Heliophorus* and *Hydroporus beetles* (1989) and (c) multiple life cycle occurs in species with fast development as in *Aedes spp.* and *Culex pipiens.* Besides, several diverse types of water bodies collectively called ``Container habitat'' exist, these water retaining structures being termed *phytotelmata* or *phytolem habitats.* The concentration and relative abundance of dissolved substances in temporary water vary more than in permanent waters, resulting in a change of water chemistry. In general, temporary waters are classified into *microhabitats, mesohabitats, and macrohabitats.* In general, microhabitats include axils of leaves and the like, mesohabitats, mainly temporary ponds and intermittent streams and macorhabitats, drying lakes and the like.

INSECTS IN ODD LOTIC ENVIRONMENTS

Melt water streams in the Himalaya are all torrents that differ fundamentally from other hill streams in India, such as those on the Western Ghats, in being exclusively glacial in origin and not rain0fed. The bottom of these torrents is also stony, with little sand or mud. All the turbulent and often run in great cascades and have many high waterfalls in their meandering journery down the valleys. The fall in gradient in most streams is high, often as much as 500-200 meters per kilometer. There are two types of associations in most of these streams viz., associations of mayflies like *Baetis, Baetiella, Ephemerella, Torleya, Hyrtanella, Epeorus, Iron and Ecdyonurus* and stoneflies like *Nemoura, Capnia, Neoperla and Phanoperla* as well as a second type of association mainly composed of Blepharoceridae-Deuterophelbiidae-Simulidae. The former type seems to avoid extreme force of the current and prefers to remain under and in between submerged stones. The latter type occurs exposed to the full force of the current. The Ephemeroptera-Plecoptera associations are abundant up to an elevation of nearly 4000 meters, but above this altitude Plecoptera begin to dominate. Diptera and Trichoptera nearly to the permanent snow-line (Mani, 1962). Principal specializations of the torrential communities include their ability not only to withstand but also to utilize the current, their organs of anchorage, absence of swift movements, small size, stremalining of body, habit of filtering from the water current or scarping food from the stone surface, stenothermy and low tolerance to oxygen deficiency.

Thermal springs are another odd environment to which some aquatic insects are well adapted. Thermal or hot springs at elevations above the timer-line are characterized by temperature higher than the mean annual atomospheric temperature of the locality. The temperature of the water inhabited by insects ranges from 30°C to 51°C. The larvae of the fly, *Ephydra* occur in the Phuga Hot spring at an elevation of 4,750 meters, at a temperature of 49.1°C (Mani, 1962). Other characteristic species of the hot spring communities belong to

Chironomidae, Ceratopogonidae and Stratiomyiidae. The larvae of these flies occur at temperatures of nearly 48°C. Small numbers of the larvae of *Eristalsis* (Syrphidae) are also ofter found in some hot spring communities. Hydrophilid beetles, *Oechthebius* and *Helophorus* are found in hot springs in Himalaya (Mani, 1962).

Ephemeral or intermittent streams occur throughout the world. Downstream movement from perential upstream sources after a dry spell and oviposition by flying insects, originating from perennial waters along with drift are the most important means of recolonizing in tropical, ephemeral streams. Depending on their location, size and rainfall patterns, these temporary habitats may persist for a few days or for months before drying out. Two distinct strategies enable insects, including chironomids, to exploit these unpredictable habitats. They may either be drought-resisting and able to survive at least some degree of desiccation (e.g.) *Polypedium* sp. or they may be opportunistic with persistent populations in permanently wet or perennial aquatic habitats (eg) *Cloeon* sp.

Stream riffles and pools

A riffle is the turbulent part of the stream that is shallow and has comparatively fast current whereas pool is deep and has a comparatively slow current, the water surface being calm unless disturbed by wind. Each has a characteristic insect assemblage. Rocky bottom specialists as well as leaf and algal shredders mainly inhabit riffles. *Baetis, Eperous,* (Ephemeroptera), *Neoperla* (Plecoptera), *Stenopsyche*, *Hydropsyche,* calamoceratids and lemidostomatids (*Trichoptera),* Gyrinus, *Dineutes,* (whirligig beetles), *Psephenus* (riffle beetle), *Stenelmis* (Coleoptera), gerrids and veliids (Heteroptera), simulids and blepharocerids (Diptera), pyralid moth, *Aulocodes* (Lepidoptera) and meglopterans (dobsonflies) commonly occur in stream riffles. They have several strategies to live in odd environments. For instance, the only truly aquatic moth, *Aulocodes* makes a tent, a grayish white sheet (1-2 cm) in diameter over a rock crevice. The larva undulates slightly to circulate water through the small holes at the tent edges and past the dense gill tufts lining the abdomen. In fastest running water, the blackfly (Simulidae) larva attaches itself to a rock or branch with a circular row of hooks at the end of its abdomen. From this anchor, it gathers food by extending a pair of retractile fans from its head that trap drifting organic particles. Blackflies achieve impressive densities very ofter in streams. Hydropsychid larval retreats are adaptations for filter feeding. The largest riffle dwellers, the megalopterans, reach upto 5 cm long. They live under stones in fast flowing rocky areas. They are obvious predators with huge jaws that are as big as their heads. Their bodies are soft, however, with long tube-like gills extending straight out from the sides of each abdominal segment and a pair of hooks at the tail end for holding on in the current. *Centroptilum, Caenia* and *Neurocaenis (*mayflies). *Helicopsyches* (caddisfly), and myriads of predatory libellulids, gomphids and corduligasterids (dragonflies) and caenagrionids and calopterygids (damselflies), predaceous diving beetles (dytiscidae) and scavenger beetles (Hydriophilidae)and heteropterans like nepids, belostomatids, notonectids and corixids are common in stream pools. Naucorids and Gelastgocorids bugs are found in stream edges. Gerrids skate on surface film, of water, with modified leg tips preventing these insects from breaking the water's surface tension. Fig. 11.1. Some insects are specialized to live as neustons, glued to surface film of water. Gyrinids have a split view of the neustonic feeding area *predators* below. Burrowing mayflies, (*Ephemera, Rhoenanthus,* and *Eatonigenia*) in the sandy bottom of streams have tusks and/or other adaptations to suit that habitat.

Hyporheos, a bizarre habitat

Probably, the most bizarre of stream habitats is the hyporheic zone, which literally means ``below the flow". Aquatic insects and other macroinvertebrates may penetrate a considerable distance vertically and laterally in hill streams with loosely packed substrata of stones or coarse gravel giving rise to hyporheic fauna. Occasional hyporheos consist mostly of aquatic insect larvae, which use this zone as a refuge during their early development, while permanent hyporheos are specialized groups that go through their entire life cycle there (Williams, 1974), Tipulid larvae burrow just a few inches under the stream bed. They use peristaltic motion to move through sand in search of food and out of water, the worm-like larvae usually have swollen posterior ends. Hyporheic fauna survice on the bacteria and fundi which live deep within the porous sand/gravel substrate where nutrients seep into the ground water enabling the entomofauna their to thrive even while the surface dries totally during the dry season. Hyporpheic zone is, indeed, an active ecotone between surface stream and ground water (Boulton *et al.,* 1998). And the invertebrate communities there can provide unique insights into ther linkage between rivers and their adjacent floor plains, ultimately playing an important role as indicators of water quality and ecosystem health.

The shredding caddisfly larva, *Anisocentropus* lives in a tunnel between two fallen leaves of riparian angiosperms, glued together nicely. Shredding insects dominate in shady, montane headwaters, where allochthonous food source is abundant due to thick canopy of riparian vegetation. Characteristic assemblages of aquatic insects are associated with submerged wood. Several species of chironomids and larvae of some species of mayflies (e.g. *Povilla*) are xylophagous (wood eating). Chironomidae and Tipulidae are invariably dominant in both abundance and diversity, burrowing and tunneling into the wood (Anderson, 1984).

INSECTS IN ODD LENTIC ENVIRONMENTS

Hazardous life in ponds, ditches, marshes and swamps

Life is insects in small, ephemeral lentic water bodies, especially in the tropics, under the spell of monsoonal vicissitudes and unpredictable rainfall patterns are subjected to frequent drying and hypoxial conditions with the result that it involves remarkable adaptability in terms of respiration, osmoregulation and reproduction. These odd environments create very different restraints from their comparatively problem free terrestrial counterparts. Oxygen is obtained by aquatic insects either by taking that available in abundance above the water or by uptake from oxygen in solution in the surrounding water. Midge larvae generate respiratory pigment, dissolved in the plasma in abundance to the extent of naming them as ``blood worms''. Tracheated cuticular outgrowths in the form of tracheal gills and rectal gills are evolved in mayflies and odonates respectively. In contrast to gilled insects, aeropneustic insects (larval Culicidae) use respiratory siphons to gain direct connection to a source of air. This may be through the water surface as seen in many Culicidae or by tapping into specific aquatic plant aerenchyma tissue, as seen in *Mansonia* larvae. Some aquatic adult beetles respire taking air bubbles which adhere to pleura due to hydrofugic hairs. This strategy is termed `*plastron respiration*'. Water scorpions too have long respiratory siphons which are modified caudal filaments. Water surrounding the insect will become deplected especially in lentic (still) waters and so many insects direct a flow of water over their bodies. *Elmis* is a lentic insect which employs a passive current control by suitable orientation. Aquatic insects inhabiting soft sediments have different strategies to withstand hypoxial conditions. For instance, some chironomids may raise the openings of their tubes above the level of sediment in which they are living. Sediment depth has also a major influence on population density and tube shape in *Chironomus plumosus*. In response to low oxygen concentrations, some caddisfly larvae undulate their abdomen, some stonefly and damselfly nymphs do ``push ups" and some mayfly nymphs beat their tracheal gills.

The internal mileu of aquatic insects in hypertonic to their environment and water passes through the cuticle and into the insects. As an osmoregulatory strategy, salt absorption occurs in the gut and Na and Ca ions are reabsorbed by an active uptake process within the rectum. Certain larvae are able to uptake salts from very dilute solutions using active processes. This can be seen in *Aedes, Culex and Chironomus,* which use anal papillae as the uptake site. Chloride epithelia are used for ion uptake in Odonata and Diptera. These are localized areas of integument, which actively take up and replace ions from the surrounding water. Insects that live in temporary pools generally have, in addition to the special anatomical, physiological adaptations some behavioural character-istics that adapt them to life in these changeable habitats. For example, if the pool dries up, some dragonfly nymphs can survive for months by burrowing into the moist bottom sediments and becoming dormant. Others leave the drying pool in the adult stage to fly off in search of temporary pools that still hold water. Water striders give no overt response to the unseen predators. However, they respond very energetically to predators like surface feeding fish and warblers that swoop down from mangroves growing in shallows, if they see them before being attacked. With early warming, the members of large groups remain aggregated, but initiate behaviour changes which are likely to confuse predators by transforming a group of slow-moving, oriented individuals into a collection of rapid and randomly moving ones. But, under similar circumstances, small groups of water striders suddenly move away from the scene when a predator comes close, — a confusing last-minute explosive dispersal (Waldbauer, 2006).

Aquatic insects least tolerant and pollution include *Glossosoma* (Trichoptera), *Rhagovelia* (Heteroptera), *Euphaea* (Odonata), *Psephenis* (Coleoptera), *Blepharocera* (Diptera), *Teloganodes* (Ephemeroptera). Moderately tolerant are: *Simulium* (Diptera), *Micronecta* (Heteroptera), *Limnometra* (Heteroptera) and *Cybister* (Coleoptera). Highly tolerant of pollution are: *Caenis* (Ephemeroptera) and *Chironomus* (Diptera).

Aquatic bugs like mesovelids (*Mesovelia horvathi* and *M. vittegra*), hydrometrids (*Hydrometra butleri* and *H. greeni*), microvelids (*Microvelia diluta, M. douglasi),* gerrids (*Neogerris parvula, Limnogonus nitidus* and *Limnometra fluviorum*), belostomatids (*Diplonychus annulatus, D. molestus* and *D. rusticus*), corixids (*Agraptocorixa hyalinipennis),* micronectids (*Micronecta scutellaris, M. quadristrigata*) notonectids (*Anisops campbelli, A. cavifrons* and *A. nivea),* pleids (*Paraplea frantalisea*), aquatic beetles like dytiscids (*Cybister limbatus,* and *Hydaticus fabrici*), dryopids (*Helichus* sp.) hydrophilids (*Hydrophilus olivaceous*, and *Stenolophus rufipes*) and many species of culicines and Chironomidae dominate the scene in intermittent and ephemeral ponds of peninsular India.

Insect communities in high altitude ponds and lakes

With the notable exception of Collembola, insect diversity is low in this habitat. Species belonging to Dytiscidae (*Agabus, Potamonectes* and *Dytiscus*) and Hydrophilidae (*Helophorus*) inhabit these habitats in Himalayan ponds and lakes. Absence of the surface haunting aquatic Heteroptera is conspicuous, with the exception of *Gerris sahlbergi* in some isolated patches (Mani, 1962).

Leaf mining and phytophagy

A number of chironomid larvae belonging to Orthocladinae and Chironominae mine in plant tissues. Larvae of curculionid weevil, *Bagous* makes tunnels while mining in leaves of the aquatic weed, *Ottelia alismoides.* Some insects are obligate phytophages, which feed mainly on living plant tissue. Others are *seston feeders*, which filter feed in mines and can also consume plant tissue. Yet others are *facultative phytophages.* Which use the plant tissue only for making mines (Van der Velde and Hiddink, 1987).

``Processing chain commensalisms" in tree holes and pitcher plants

`Plant waters' or ``Phytotelmata" are common in the tropics and are classified on the basis of their position in the plant, containing predominantly rain water. They are common in plants belonging to Araceae, Bromeliaceae, Cannaceae and Pandanaceae (Kitching, 1971). Larval mosquitoes frequently occur in the axil waters. Water reservoirs are found in the tubular stems of bamboos, the faundal complex comprising culcids. Chironomids, ceratopogonids and a few species of Odonata. By far, the most common type of phytotelmata is the water filled tree hole, both in fallen, horizontal trees, as well as upright living trees. As such tree holes form an important part of the aquatic environment of insects. The size of the tree hole and species of trees have a bearing on the mosquito fauna. Broadly, two kinds of tree holes are recognizable - `*pans,* which have an unbroken bark lining and *rot holes,* penetrating their wood tissues (Williams, 1987). Since,

Fig. 11.1. Temporary water bodies

Fig. 11.2. Temporary water bodies

Fig. 11.3. Temporary water bodies

Fig. 11.4. Tree holes

Fig. 11.5. Tree holes

Fig. 11.6

rot holes contains both litter and fungus-rotten wood, insects such as mosquitoes, chironomids, biting midges and mould beetles like helodids and scirtids thrive there. Predatory tree hole mosquito, *Toxorhynchites* her a comparatively long life span, the females laying 1-7 eggs in tree holes. Species in three holes are broadly divisible into `*dendrolimnetoxene'* (accidental inhabitants), `*dendrolimnetophile'* (those attracted to any semi-aquatic moist habitat) and `*dendrolimnetobionten'* (species specially associated with water-filled tree holes such as species of *Aedes, Anopheles, syrphids,* chironomids, ceratopogonids, Coleoptera such as heliodids).

Water filled tree holes can be considered as semipermanent habitats, with the susceptibility of drying out. The fauna of tree holes have the ability to withstand considerable period of dryness at least in some stage of their life history (Kitching, 1971). Bottom of tree holes remain moist, enabling the inhabitants to survive. Furthermore, tree holes are essentially larval habitats and on emergence, the adults are dispersed throughout the woodland. In a nutshell, water-filled tree holes are `discrete habitats' (Kitching, 1969), with the ability to resist extremes and rapid variations of environmental conditions, combined with their ability to multiply rapidly, in view of the short periods of favourable conditions.

Tree hole communities are discrete and relatively small subsets of the forest ecosystem, which makes them very useful for studying processes and patterns occurring at the level of the community. Two types of common insect larvae in tree holes, scirtid beetles and ceratopogonid midges have a commensalistic relationship (Paradise and Dunson, 1997). Scirtids are leaf shredders, skeletonizing leaf litter (Barrera, 1988). A portion of what they shred is nost consumed by them and consequently made available to insects and microbes feeding on detritus. Midges benefit from this by growing to a large size and having a higher survival in the presence of scirtids. However, the presence of midges do not affect scirtids. This kind of ``*processing chain commensalism"* may be general feature of tree hole-like communities (phytotelmata) as it has also been observed in pitcher-plants (Bradshaw, 1983; Heard, 1994a) and *Heliconia* bracts (Seifert and Seifert, 1976; 1979). These assemblages constitute detritus-based food webs and organisms that consume detritus in different forms would be more likely to interact under conditions of resource limitations (Heard, 1994a; 1994b). In fact, the interaction between scirtids and midges is more pronounced at low resource levels (Paradise and Dunson, 1997).

Variation in food webs both spatially and temporally and are common features of phytotelmata (Kitching and Beaver, 1990). Spatial heterogeneity should be examined across climatic zones, regions within zones and at local scales within ecosystem. At the local level, this variability reflects stochasticity of certain processes, such as search behaviour of ovipositing females and heterogeneity of abiotic factors and resources (Kitching and Beaver, 1990). Tree hole communities are used as a model system to study the interplay of abiotic and biotic factors, the impacts of atmospheric deposition and the biotic interactions among insect larvae inhabiting a detritus-based community. There are similarities to other communities that allow tree holes to be used for the study of these general ecological questions. Although this is a small, discrete habitat, its complexity is obvious. This also makes it realistic as a model for larger scale communities. There are many relationships among abiotic and biotic factors worth exploring in future in three holes and other similar habitats. The relationships among abiotic and biotic factors in tree holes from areas with different deposition patterns need further investigations.

Insects in intertidal zones, estuaries and backwaters - life with severe constraints

Though insects have achieved unparalleled success in both terrestrial and freshwater environments, they are not well represented in the sea, as the number of species that can be called `marine' amounts to only seven hundred (Williams and Williams, 1998). Past evolutionary history reveals that, much earlier to the arrival of the primarily terrestrial taxon. Insecta on the marine scene, practically all conceivable niches there were occupied mainly by crustaceans and also by molluscs. Apart from the hemipteran genus, *Halobates*, no insects are known from the open ocean. Undeniably, the majority of insects that have been recorded from marine situations have been collected in so-called `bridging habitats' eg. estuaries, salt marshes, interidal areas, mangrove swamps etc. (Cheng, 1976). However, there are a few extremely examples of aquatic insects well adapted to a marine existence. As mentioned earlier, the gerrid genus, *Halobates*, comprises some species that live on the surface of the oceans, especially in the Indomalaysian and Australian region. Five of these live in the central pacific Ocean and have no contact with land (Cheng, 1985). The chironomid, *Pontomyia,* which completes its life cycle at depths of 30 m on tropical coral reefs (Bretscko, 1982). Besides this, a number of chironomid species are tolerant of a wide range of salinities and are a major component of the fauna of brackish waters. Though salt water environment is a major constraint, the ability of many aquatic insects to tolerate inundation by saltwater attests to their changing osmotic, physiological and possibly, behavioural capabilities. Perhaps there are genotypic differences in the abilities of subpopulations of these freshwater insects to cope ith varying salinity. This aspect needs more precise investigation in future?

``ECOSYSTEM ENGINEERING'' IN ODD ENVIRONMENTS

Adaptive ingenuity of aquatic insects in some odd stream habitats is amply illustrated by the ``ecosystem engineering'' activities by some mayflies, stoneflies and caddisflies in terms of habitat creation, bioturbation, bioconsolidation and particulate matter processing. For instance, caddisflies physically modify habitat by consolidating benthic sediments.

Hydropsychids play and important role in generating the spatial ``regues'' that moderate resistance of lotic communities to flooding (Cardinale *et al.,* 2004). The presence of stoneflies enhances sediment removal from interstitial spaces, thus impacting benthic microhabitat quality (Zenetell and Peckarsky, 1996).

INSECTS IN ODD ZOONOTIC ENVIRONMENTS

Many aquatic insects are associated with other macroinvertebrates by lodging themselves in their body parts. Two interesting associations of baetid mayfly nymphs with freshwater clams (mussels) were recorded previously, one in Thailand (Muller-Liebenau and Heard, 1979) and the other in Mali and Guinea in West Africa (Gillies and Elouard, 1990). The former association is between the mayfly, *Symbiocloeon heardi* and the clam, *Hyriopsis myersiana,* whereas the latter association is between the mayfly, *Muteloclοeon bihoumi* and large mussels belonging to the general *Mutela, Aspartharia* and *Spathopsis* (Mutelidae). Subramanian and Sivaramakrishnan have recently recorded a similar mayfly-mussel association from Tunga river in Western Ghats of India. This association may be regarded as one of commensalism, beneficial to the insect, neutral to the mollusk (Muller-Liebenau and Heard, 1979), although in view of the internal lodgement of the nymph among the fimbriated lamellae of the gills, the mayfly might well be regarded as an inquiline (Gillies and Elouard, 1990).

CHAPTER - 12

URBAN ENTOMOLOGY : INSECT ECOLOGY IN URBAN HABITATS

Introduction

The urban ecosystem comprises man-made elements comprised of both living and non-living components influenced by man, essentially involving housing as well as transport networks, with plants and animals occupying specific habitats. While agroecosystems are essentially created and fostered by man the urban ecosystem represents ``the highest expression of human technological development and cultural evolution" (Robinson, 1996). To further quote Robinson, ``Urban populations are growing by about 3.8% every year, and by 2020, it is expected that about 4 million people will inhabit urban areas ... megacities with more than ten million people and 13 of the megacities will be in Asia". Naturally the variety of pollutants and health hazard tend to be phenomenal, accompanied by diverse pestiferous insects which tend to enjoy favourable conditions, not to mention of the diversity of plant life and associated organisms. Needless to indicate that the absence of competition for food resources has enabled them to gain a strong foothold in the urban environment. Peridomestic populations existing outside human habitats, such as beetles, bugs, caterpillars feeding on ornamental plants, besides cockroaches and other arthropods such as stored grain pests, flies, carpet beetles and dust mites which exist indoors tend to affect human health and property, some acting as vector of disease. The emergence of household gardens, parks as well as roadside avenue trees has also contributed to the emergence of several pest species. In short, while *agrobocoenosis* involves the occasional utilization of small plots of land in urban environments, *anthropobiocoenosis* includes human habitats including buildings and structures associated with such habitats. Turf grass, ornamental plants and the like, grown by man and the organisms associated with the domestic environment including termites, ants, wood boring beetles, bees and warp are included urban anthropobiocoenosis.

Increasing environmental onslaught resulting from ever growing demands for food, housing, power, goods and personal mobility, thanks to ever increasing population build-up, has resulted in increasing urbanization, the consequences of which are being realized today. In urban centres, man virtually has suppressed all life that ever existed there, substituting a community of little variety that repeats itself. Instead of soil, natural water system plants and animal communities, there exist concrete surfaces, sewage pipes and parks with alien and animals. Land use changes, land cover result in major changes in the diversity of species member and biomass of anthropods. A such every technical advance brings with it ecological hazard. Francis Bacon has said ``We cannot command nature, except by obeying her''. The highly modified nature of the urban ecosystems tend to make them especially vulnerable to alien species invasion (Gilbert, 1989). With extensive lawn coming into being with increased urbanization, one of the major changes relates to the diversity of ant species the control of which has become a major problem. With every passing day more and more lands continue to go in for housing projects and lowlands adjoining swamps have become a handy repository for garbage of all sorts. As such urban development leads to several kinds of habitat loss including qualitation and quantitons Hanski (2005) indicates four kinds of losses - the loss of quality, quantity (area), connectivity (reduced immigrates and colonization ration) and of continuity. Loss of microhabitats go hand in hand with macrohabitats. Intensive landuse for purposes of urban development has resulted in structural simplification of several macrohabitats. Parks, urban forests and gardens are cleared of woody and leafy debris, and extensive lawns replace more natural grassland eliminating microhabitats near the soil surface (Gilbert, 1989). In today's scenario loss of wetlands, swamps, marshes and shallow lakes, are dominated by grasses, sedges and reeds. Water surface is covered by scum of green algae, besides the large scale occurrence of water hyacinth with its bluish bloom and loose mat of floating leaves and fine filamentous rooli housing all sorts of small organisms. Insect life abound in such swampy areas presents an enormous diversity of shape, size and colour, teeming with predators, aquatic insects besides others, presenting a microcosm of insects.

Urbanization is one of the leading causes of species extinction, habitat alteration tending to be drastic and increasingly widespread, usually preceded by widespread destruction of much of native vegetation. Urbanization has had a drastic effect on the abiotic components of environment, especially in larger cities. Urban adaptabler species become increasingly widespread and locally abundant in cities. With regard to insects, urbanization has found many new habitats and modified existing ones. Numerous insects are associated with wastes accumulated by man such on garbage dumps, sewage outfalls and drainage ditches providing suitable habitats for insects. With the expansion of cities biological homogenization increases because the smae urban- adaptable species becomes increasingly widespread and locally abundant in cities (2006). The maximum disturbance created by growth of cities not only destroy the habitat of the local species, but also create a habitat for relatively few species that become adapted to urban and semiurban conditions. The construction and expansion of cities and towns promote the loss of native species and their replacement by non-natve species that did not occur in the specified area. For instance, with the removal of large scale plantation of *Acacia* and *Prosopis* to make way for housing purposes, some key rare species of insect as the viviparous roack, *Trichoblatta sericea* common under the bark of these trees have become scarce. In general, urban activities find to cause non-native species to increase in numbers and native species tend to decline. The once widespread species of *Chrysocoris* abundant on the weed *Croton sparsiflorus* has now become rare.

Urban ecosystems are quite similar world wide in terms of structure and functional constraints. Population density, fragmentation of natural vegetation, air, soil and water pollution, higher ambient temperature than surrounding areas, increasing alkalinity of soils, promoting plants which require high pH, with natural accompanying changes of soil fauna. Urban heat island effect provides a microclimate that is much warmer than the surrounding landscape. Species vary widely in their ability to adapt to drastic changes along the urban-rural gradient. The responses along the urban-rural gradient are *avoidance, adoption and exploitation.* Some species being extrasensitive disappear quickly (avoidance), some thrive to the point where they become adapted to urban habitats and also able to utilize natural resources (adaptation), with the exploiters becoming dependent on urban resources. In short they are *urbanophiles* (positive) and *urbanophobes* (negative) regarding responses to urbanization. Exploiters are urbanophiles, while adapters are urbanophobes. Exploiters are inversely correlated with vegetation and in contrast, adaptation requires considerable vegetation for shelter and food. As such biotic and abiotic factors deterring population abundance of each species along the urban-rural gradient are often complex and tend to vary with species. Urbanization leads to high floristic diversity in parks, gardens, residential landscapes, and consequent changes in insect diversity. As such the process of urbanization has resulted in the expansion of alien plant species and decline of native species causing greater similarly between different urban areas leading to *biotic homogenization.* Therefore urbanization acts in diverse ways on existing bioloy, by altering the quality of air, water, soil, temperature, rainfall pattern, habitats fragments and disturbance. Urban environment therefore has been considerably influenced by human activities, with a pronounced effect on the abiotic components. Urbanization has therefore resulted in new insect habitats, with many of them exploiting fences, wooden shrunken, the latter being favourable habitats for ants and termites, Cockroach diversity in relation to human surrounding has equally been considerable, with species of *Supella, Blatta,* besides mosquitos in flower pots, glass jars, cans and garbage dumps.

As such many insects have established ecological relationships with some species dispersing naturally to new areas where man has established their host plants enabling oviposition in new sites. Aphid epidemics occur primarly in urban areas, not to mention of white flies. In short, insects tend to disperse from ecologically young, i.e. more recent environments, to ecologically old areas. The diversity of wasps such as mud wasps, pot wasps, paper wasps, not to mention of ants can easily be anticipated. With insecticides being part of the operational environment, selection tends to favour those that can resist the compounds, such as biological differences between resistant and susceptible house flies and mosquitoes. *Aedes aegyptii* has adapted to the urban environment with two or three forum viz. domestic, predomestic and feral, with the domestic form being twice as resistant as feral ones. Urbanization can therefore enhance survival of both pathoform and verton through suitable habitats by providing increasing possibility of transmission of a pathogen in high density areas. Human activities in short, have feed the species from competation or natural enemies. Ecological aspects of urban vegetation reflects some socio-economic features, with the community being a part of the cultural environment (Gilbert, 1989). Insect pollination is central to the maintenance of plant diversity in urban areas and reduction in natural pollination because of pollutants tend to be disastrous.

The proliferation of introduced species in urban environments impose considerable ``inoculation pressure'' (Gilbert, 1989) or urban habitats, with in course of time, discovering vacant miches in which to become established, such as these of herbivorous plants, with increased opportunities for dispersal. While Labiatae, Asteraceae are very common, those with the capability to exploit favourable man-made habitats are Asteracae, Labiatae, Leguminoseae, Orchidaceae an Graminae. As vegetation changes with succession phytophagous insect fauna change, generalist or polyphagous species tending to be replaced by specialist feeders their distribution being controlled by the four plants. Another specialist feeder pertains to *Poekilocerus pictus* in urban environments, notably occuring on *Calotropis* in vacant sites and along railway tracks. Similarly beetles, notably carabids are common in demolished sites. Further, the changing face of transport and industry results in changes in faunal resources such as the oil milling industry pertaining to *Amaranthus, Helianthus, Hordeum, Sesamum* etc. not to mention of pests occurring in the paper milling industry and cotton industry. A notable example in the early urbanization periods is the profuse growth of *Croton sparsiflorus* infested with large scale occurrence of *Chrysocoris* or jewel bugs.

Urban areas are characterized by high levels of disturbance and environmental modification so that a of knowledge vegetation dynamics is basic to our understanding of ecological interactions. Besides there is increasing loss of herbivore communities and replacement initially by bare earth, followed by grasses. Reduction in the biomass of trees and shrubs and the loss of biodiversity through severe structural modifications coupled with fragmentation, resulting in course of time, in vegetation being no longer related to environmental vaguaries. As vegetation changes, with succession, so does phytophagous insects with their niche breadth declining as succession proceeds and in course of time generalists tend to replace specialists. The broad patterns of arthropod invasion being : -ants - butterflies - carabids Hemiptera-Orthoptera and bees. Small groups of pollinators favour certain groups such as psocids like *Mesopsocus, Reuterella* which prefer feeding on the alga *Pleurococcus,* not to mention of collembolan and thrips. In polluted cities or urban environments, the impact is great on epiphytes which tend to become eliminated and consequent insect associates such as psocids, lichenophagous caterpillars, etc.

Scientific research in the past form decades has shown how much civilization itself depends on the existence of healthy wetlands. Swamps store water with much less evaporation than reservoirs, swamps and marshes purifying sewage effluents more easily, so that wetlands are not simply living museums for the preservation of

Fig. 12.1. Land fills as a prelude to Urbanization

Fig. 12.2. Damage to Jamun fruits by *Mallothrips indicus*

crocodiles ``but a vital part of the hydraulic system that produces clean and reliable water supplies. In today's scenario, wetland and marshes have disappeared, being invaded by garbage dumps which gradually replace them. One of the notable examples relates to the Pallikkaranai marsh, the last of the wetlands bordering Chennai, which is on the verge of disappearance, thanks to the largescale urbanization, and is one the major dumping yards. Being low lying and a perennial water source with an extensive catschment. It serves as a drain for flood water during monsoon. Increasing construction activities and garbage dumping tend to destroy the fragile ecology of the region (Oppili, 2001). Such human interference and construction activities have reduced the marsh to a smaller area (Fig. 12.1).

As such special ecological relationships exist in relation to temporary ponds, lakes, canals and similar water bodies. Water collected in gutters and drains are discharged into the nearest water course, as for instance the Buckingham Canal in Chennai Meteropollitan area. Alternate periods of heavy flow, followed by a drying out often occurs. A wide range of pollutants are discharged as water passes through urban areas which considerably affects its ecology. These water bodies become densely clothed with bacteria, fungi and predatory protozoa, absorbing most of the O_2 from water causing conditions in which very few organisms thrive. Eutrophication in lakes or rivers promotes alagal growth resulting in green water effluents with algae like *Cladophora, Spirogyra, Ulothrix* which cover the entire surface with a greenish mat. Besdes, industrial effluents contaminants are also toxic. In short, water quality in the urban environment is largely controlled by the efficiency of sewage treatment. Aquatic bryophytes can function as useful indication of pollution. Mosses are highly tolerant of heavy metal pollution, while the algae like *Cladophora* are sensitive. Some of the insect groups showing increasing sensitivity to water pollution are chironomids, baetids, caddisflies, *Mesovelia*, hydrometrids, gerrids, *Laccotrephes*, corixids; beetles like Haliplidae, dytiscids, gyrinids, caddisflies, and stoneflies; may flies like *Leptophlebia, Heptagenia, Ephemerella*, show increasing sensitivity to water pollution.

Urbanization has profound effects on the abiotic components of the environment, involving aspects such a heat absorbing surfaces of roads, limited amount of green space and open soil, large amounts of pollution and particulate matter in the air. Urban areas as a result have different climate from the surrounding country side. The heat emanating from the industrial smoke, vehicles and machinery result in urban heat island. As such the urban ecosystem has two sectors, one interfacing with the suburban landscape with 60% green space with a range of soil types, ground cover, plant and animal species; the other sector is the most developed ecosystem with over 60% of the area consisting of a hard surface and built structure with an uneven distribution of exposed soil and sparse vegetation (Robinson, 2005). Availability of flying insects including flies, beetles, plant bugs, moths are attracted by street lights, more especially of the large numbers of the giant water bug *Lethocerus indicus* soon after rains. In the urban environment, honey bees are members of the urban insect community, manipulated to provide honey, bees wax and pollination services. Moving into urban areas are the nests of birds, paper nests of social wasps and bees, these are constructed close to human dwellings. The carpet beetle *Anthrenus* common in wasps nests and the drug store beetle *Stegobium* is equally common in bees nests. Silver fish and fire brats have also been reported moving into the household environment from nearby birds nests (Robinson, 1996).

Insects attracting food and fabric from outside the domestic environment have become most destructive, such as the clothes moth, *Tineola bisellliella* and the Indian meal moth *Plodia interpunctella* feeding on flour and corn. Several ant species are successful in making the transition to the urban environment, getting adapted to soil types and vegetation. The abundance of turf grass and ornamental trees and shrubs in the urban environment facilitate the transition of many ant species from the agricultural environments. Fly species tolerant of changing habitats and those with a range of larval feeding habits, have been the most successful insects to adapt to human environment. Female flies that moved away from their original breeding habitat, find suitable egg laying substrates associated with human habitats viz. dump sites of household garbage and accumulated of human and animal waste which are excellent sites for egg laying. The mosquito *Culex tritaeniorhychus* has successfully adapted to the human environment and rapid urbanization has helped the spread of *Aedes aegypti*.

``The large number of foreign organisms that are intentionally and unintentionally introduced into cities each year, impose a considerable inoculation pressure on urban habitats, with the result, that many of the species eventually find a vacant niche and become established" (Gilbert, 1989). The establishment of gardens both around houses and parks is a haven for many insects. For example, diverse varieties of crotons, which adorn the urban environment, brings with them such insects as coccids, white flies, butterflies like *Papilio and Danaius*, aphids and thrips such as *Heliothrips haemorrhidalis*. A brief list of insects infesting ornamental plants in the urban environment is appended. Jasmine is a favourable twiner cultivated for ornamental purposes, hosts, several species of thrips. Similarly the need to assess the insect fauna of road-side avenue trees such as Ficus, Jamun, Tamarind, *Mangifera, Peltophorum*, Neem, *Pongamia*, becomes imporant for an overall assessment of the insect fauna in an urban environment (Fig. 12.2).

A brief list of some of the common insects on road-side avenue trees is provided below :

Tamarind	*Phycita orthoclina* (Fruit borer)
	Holotrichia insularis (White grub)
	Dichocrocis punctiferalis (Capsule borer)
	Aspidiotus destructor (scale)
	Virachola isocrates (Butterfly)
Jamun	*Indrabela tetraonis* (Bark-eating caterpillars)
	Holotrichia insularis (leaf beetles)
	Acrocerops phaeuspora (leaf miner)
	Carea subtilis (leaf eating caterpillar)
	Curculio album (Fruit weevil)
	Mallothrips indicus
	Polychrosis cellifera (leaf roller)
	Dialeurodes engenia (White fly)
Jak	*Perina nuda* (leaf webber)
	Batocera rufomaculata (bark borer)
	Indrabela tetraonis (Bark eating caterpillar)
	Cosmoscrata relata (spittle bug)
	Sthenia grisator (cerambycid)
	Ferrisia virgata (White tailed mealy bug)
	Ceroplastes rubous (waxy scale)
Mango	*Batocera rufomaculata* (Stem borer)
	Sternochaetus mangifera (stem weevil)
	Rhynchaenus mangiferae (leaf mining weevil)

Apoderus tranquebaricus (leaf twisting weevil)

Rastroccus iceryoides (coccid)

Amaraemyia sp (leaf galls)

Alongside introduction of diverse insect species into the urban environments, there also exist biocontrol agents like parasites and predators. Some of the major predators one comes across in urban surroundings are –

Hierodula superstitiosa	
Leptomantis indica	Mantids
Tropidomantis guttatipennis	
Orius maxidentex	
Montandoniola moraguesi	Anthocorids
Carayonocoris indicus	
Cardiastethus exiguus	
Acanthaspis siva	Reduviids
Rhinocoris marginalis	
Rodolia cardinalis	
Menochilus sp	Coccinellids
Cryptolaemus sp	
Scymnus sp	
Brachythemis sp	
Diplacodes sp	
Pantala flavescens	Dragonflies
Orthetrum sp	
Polydorus sp	
Bembidion sp	Carabids
Notophilus sp	
Cantheconidia furcellata	Pentatomid
Geocoris tricolor (lygaeid)	

Adaptability to varying conditions and changing habitats are essential aspects of urbanization and successful adaptation is evident in cockroaches, fleas, flies, ants and termites. Conditions indoor and outdoor tend to restrict the diversity of some insects. Hymenopterans like vespids and *Polistes* form single combs supended from roofs of houses, while several nests are established in soil, trees or surrounding houses in urban areas.

CHAPTER - 13

CHANGING HABITATS AND VECTOR DYNAMICS

Global climate change is likely to have important consequences on the distribution and intensity of vector borne diseases. Effects of global warming may be more significant than those of man-made environmental changes. Changes in land-use deforestation, irrigation and urbanization have profound effects on the severity of vector borne diseases. Human transition from a hunter to a farmer was important in the development of human-feeding behaviour of several vectors, as also the domestication of animals, which also became increasingly abundant hosts. Natural breaks in the forest and traditional vegetational zones being areas where vertebrates tend to congregate and often associated with aquatic habitats, water became an important environmental focus for the contact between vectors that develop in it and the hosts that regularly visit water sources for drinking. As human populations continue to increase and strip the land of its forests and also of the normal blood sources of many blood feeding vectors like mosquitoes and ticks, human-vector contact to increase with increased chances of infection.

One of the most important malarial vectors in India, *Anopheles stephensi* is adapted to breeding in overhead water tanks so that transmission in urban areas is just as intense as in the country side. *Aedes aegypti,* the dengue vector and the filarial mosquito, *Culex quinquefasciatus* and responsible for transmission of dengue and filarial respectively in urban areas. Dengue is the most important and widespread of human arboviral diseases causing high fatality rates and is entirely associated with man-made environments breeding even in flower vases, discarded car tyres, coconut shells, water storage drums and tanks, With ever increasing urbanization and associated human movement, dengue spreads at an alarming rate. The Japanese encephalitis virus resulting in brain damage could be fatal, and is transmitted by rice field breeding mosquitoes and it is a zoonosis particularly associated with birds and pigs, which act as reservoirs and amplifying hosts. Global warming will cause more intense and perhaps more prolonged transmission. Needless to mention that one of the most abundant human biting mosquitoes in tropical cities and one of the products of increased urbanization is the filarial mosquito, not to mention that one of the most abundant human biting mosquitoes in tropical cities and one of the products of increased urbanization is the filarial mosquito, not to mention of the most profound effects of spread of malarial mosquitoes as a result of water resource development, deforestation and urbanization. An aspect of importance relates to the shortening of the extrinsic cycle of the parasite as also the gonotrophic cycle of the vector, which will tend to increase the frequency of human feeding with consequent increase in transmission.

Changes in land use resulting from deforestation, irrigation and urbanization have had profound effects on the serenity of vector borne diseases. Human transition from a hunter to a farmer was an important factor in the development of human feeding behaviour of several vectors, not to mention of the domestication of animals which also became increasingly abundant hosts. Natural breaks in the forest and traditional vegetation practices have enabled vegetation to congregate become acquainted with aquate habitats. As such water became an important environmental focus for the contast between vectors that develop in it and the hosts that regularly visit these water sources for drinking with increasing urbanization the changes of human vector contrast have been on the rise, hosts increased changes of infection. One of the most important malarial vector in India *Anopheles stephensi* is adapted to breeding in overhead water tanks, so that transmission in urban areas is just as intense as it is in the ruralareas. *Aedes aegypti* the dengue vector and the filarial mosquito, *Culex quinquefasciatus* are responsible for the transmission of dengue and filarial respective in urban area. Dengue is the most important and widespread human arboviral diseases causing high fatalors rates being entirely associated with man-made environments, breed even in flower vases, discarded car tyres, coconut shells, water storage drum & tanks. The Japanese encephalitis virus spread by the vector *Culex vishnui* resulting in brain damage tend to the fatal and it transmitted for the rice-field bread mosquitoes and it is a Zoonosis particularly associated with birds and pigs which acts as reservoirs and amplifying hosts. Needles to emphasise that one of the products of increased urbanization is the filarial mosquitoes, besides equally pronounced effects of spread of malarial mosquitoes, besides equally pronounced effects of spread of malarial mosquitoes as a result of water resource development, deforestation and urbanization. While *Aedes aegypti* the yellow fever mosquito is urban, inside houses, *Aedes albopictus* is suburban or rural inhabiting forest edges and other vegetation, breeding in tree holes.

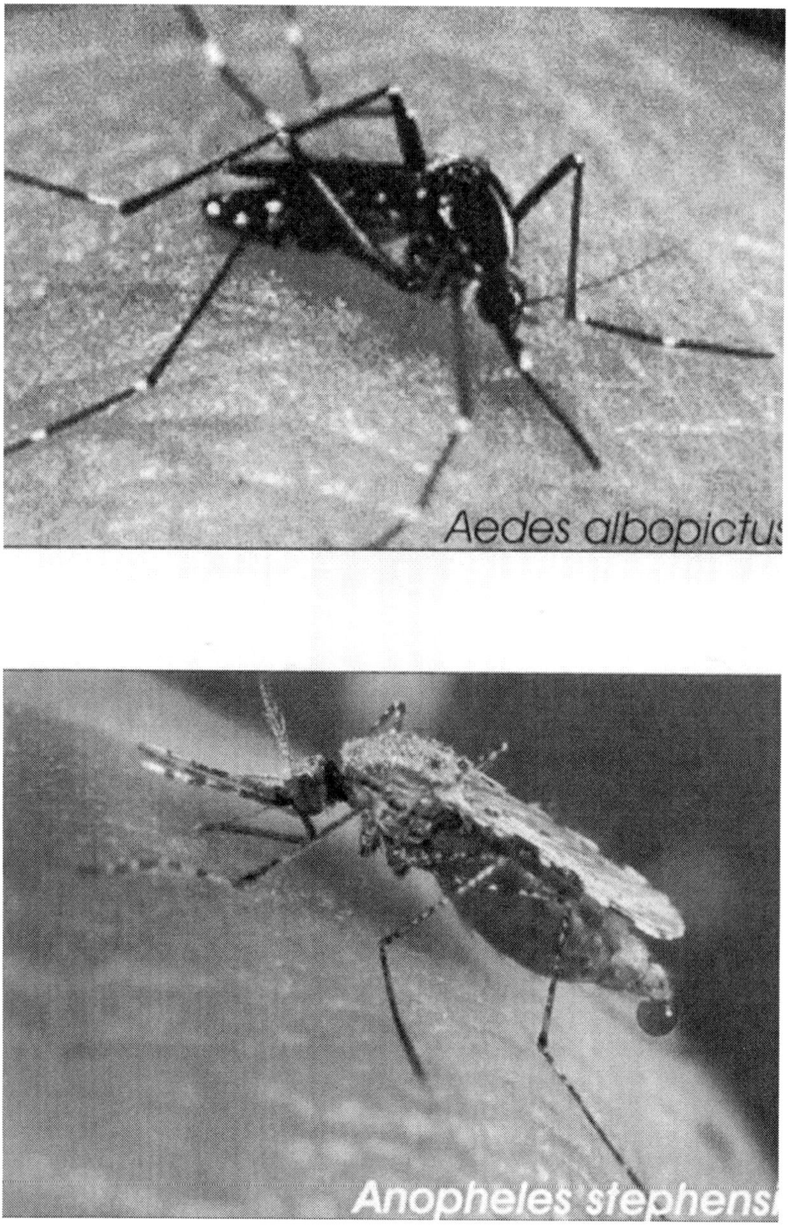

Fig. 13.1. Vector mosquitos (After Tyagi)
Aedes albopictus (Top); *Anopheles stephensi* (bottom)

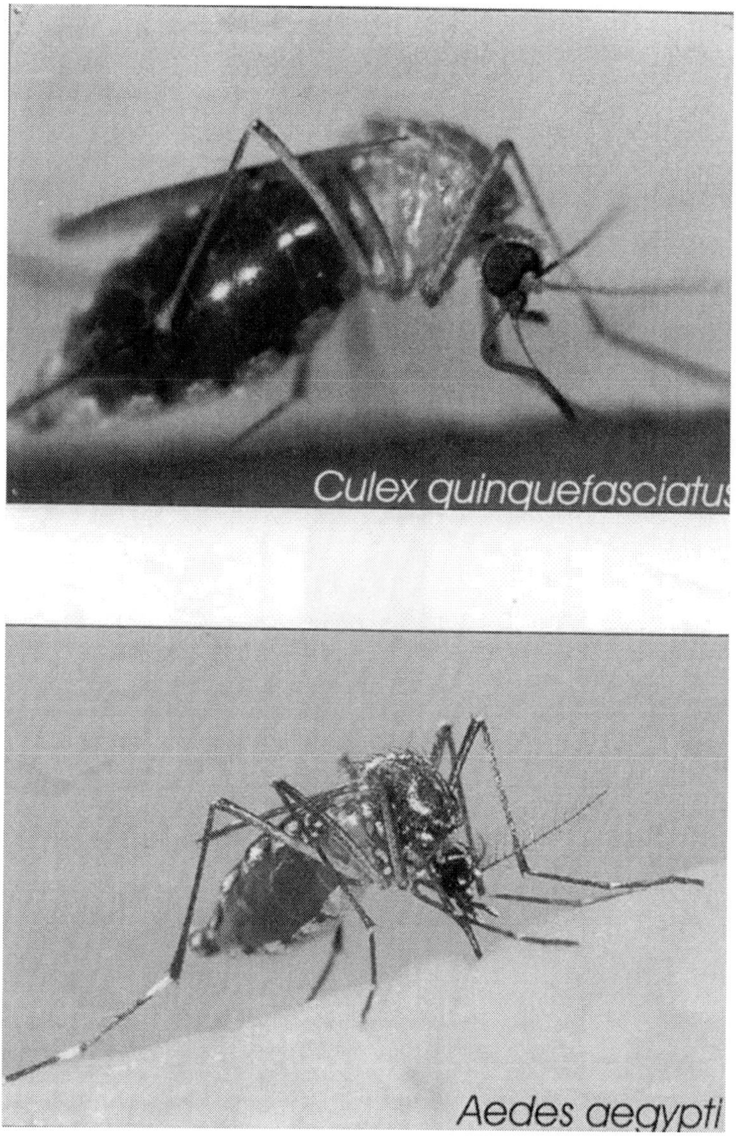

Fig. 13.2. Vector mosquitos (After Tyagi)
Culex quinquefasciatus (Top); *Aedes aegypti* (bottom)

Changed circumstances are bringing about new perceptions and demands. New approaches to forestry are being initiated, with the increased recognition of the role of carbon cycles, other nutrient cycles, water cycles etc. As a result of the increased combination of cyclical, ecological and climate processes there also results increase in biodiversity. Global rise in temperature has induced *Aedes aegypti* increase in its additional distribution by 1000 ft in the lower Himalayas. Similarly denudation of forest cover support another vector *Aedes albopictus* in the Western Ghats. No doubt, the behaviour, development and population of mosquitoes are strongly influenced by climate viz. temperature rainfall and humidity rainfall provides none breeding item and consequent increase in mosquito vector population. Vectors of Japanese encephalitis virus - *Culex tritaeniorhynchus* and *Culex psendovishnui* have a strong relationship to rainfall and particular agricultural practices. Land-based irrigation in the Thar desert has exacerbated disease scenario under the changing environmental impact through Canal-based irrigation agiculture a model par excellence for increasing of mosquito borne disease (Tyagi, 2004). At many places along the main Indira Gandhi Canal, the problem of water seepage and water logging have created serious mosquitogenic conditions, resulting in plentiful growth of vegetation, notably hydrophytic weeds providing adequate sites for mosquitoes. The greatest discernable change therefore relates to the extensive canalization resulting in stagnant water bodies providing favourable breeding sites rates for vectors of Leishmaniasis, of dengue vectors, mal and vectors, involving a number of species complexes-such as these of *A. culicifacies, A. fluviatalis, A. subpictus.*

Some human activities encouraging human bites as well as increased mosquito activities near human are :

(a) Storage of household water in clay pots

(b) Construction of open well and fish tanks

(c) Crops irrigation and water diversion

(d) Construction of pit Latrines

(e) Enrichment of surface water

(f) Impoundments in running water

(g) Improper dispersal of water holding refuse

(h) Cultivation of water holding plants.

Mosquito fauna in odd environments

The mosquito biodiversity is very rich, and is found breeding in various biotopes. The major portion of the mosquito fauna of the world is recorded from wooded areas. However, mosquito species have been recorded from various odd biotopes such as brackish water, desert, high altitudes etc. *Culex quinquefasciatus* is the species recorded throughout the tropical and subtropical regions up to altitude of 5000 ft. There are some rare species, which prefer only high altitudes eg., *Anopheles gigas, Hodgesia baileyi* (reported from India). *Anopheles sundaicus* prefers only brackish water, which is a major vector of malaria in Andaman and Nicobar Islands, and during 2004 tsunami it was expected that this species might be swept away along the shores off the main land. In mangrove forest areas of south India, *Verrallina lugubris* was found biting man extensively and was recorded breeding in crab holes along with *Rhinoskusea portonovoensis*. The fate of these two species is yet to be ascertained during post tsunami period. *Armigeres joloensis* is a rare mosquito species found breeding in tree holes in Assam, India, and in Philippines. There is a report of shift in Thar Desert where *Anopheles d'thali* also has been reported after a gap of 60 years from the Indian mainland. *Aedes albopictus* the ``Asian Tiger" mosquito of Southeast Asia entered USA nearly two decades ago and is now well established in 16 States, indicating that this species may adapt to any odd and new environment (Tyagi, Personal Communication).

Fig. 13.3. Diverse mosquito breeding grounds (After Tyagi)

As to the risk of outbreaks of vector-borne diseases, studies have shown that ground pools, puddles, coconut, groundnut and gardenpits, wells, cement lined tanks meant for water storage and water pools in and around remain of demolished sites are favourable sites for mosquito breeding. While the effect of urbanization on the availability of suitable breeding sites tend to reduce transmission rates, the main pollution factor is domestic waste water, wells and water storage tanks. The type and distribution of landscaping around houses in urban and suburban areas tend to influence the abundance of mosquitoes. The more the bees and vegetation, the greater the abundance of female mosquitoes resting in those habitats and opportunities exist for larval breeding in tree holes. Urbanization of malaria is an important aspect, the urbanization process and the resulting concentration of people and accompanying environmental pollution tend to alter the biological and vectorial factors in the transmission of malaria. As such changes in weather and climate that can affect transmission of vector borne disease include temperature, rainfall, winds, extreme flooding or drought. The natural tendency of the vector is to seek the most suitable microclimate for survival, such as resting under vegetation or pit latrine during the day. In rural rain may increase larval habitats and vector population size by creating new habitats. Incidentally, the mosquito species encountered following the Tsunami on 26th December 2004 are: *Anopheles barbirostris, Anopheles peditaeniatus, Anopheles subpictus, Anopheles vagus, Culex culicifacies, Culex tritaeniorhynchus, Culex sitiens, Aedes aegypti, Aedes vittatus.* (Courtesy - Tyagi).

CHAPTER 14

CONSERVATION FOR HEALTHIER ENVIRONMENT

Human activities in the past few centuries and particularly in the past few decades have brought the Earth to the brink of biotic crisis. The advent of the industrial revolution resulted in scant concern for the environment resulting in the gradual erosion of natural resources. Natural systems always suffer from ecological stress resulting from seasonal floods or drought, landslides and the like, but in today's scenario human activities in the past few centuries have confirmed that he is the perpetration of such changes, particularly with his role in deforestation nd associated climatic changes, misuse of land and eroding landscape, over looking environmental awareness. Heavy soil erosion has resulted in the hills due to logging and slash and burn agriculture. Perceptible scars develop in the landscape due to livestock grazing, clearing and burning. Landslides through loss of soil cover and with grazing pressure becoming heavier and heavier, the grass cover shows signs of deterioration. Deforestation removes both nutrients and principal nutrient pathways, ultimately resulting in grass changes in the abundance of soil organisms. Mycorrhizae are very common on roots of tropical forests and mycorrhizal association with tree roots play an important ecological role in forestry, their absence giving rise to symptoms of disease. Notable example is the Thar desert in Northwest India, which is a man made desert caused by extensive deforestation, overgrazing and soil deterioration. The valuable depletion of biological material through disruption of natural ecosystem is very evident. Loss of vegetational cover and soil erosion have destroyed considerable population of soil organisms so essential for the building up of soils. There is increasing evidence that community structure, species abundance and biotic interactions in relation to landscape change are tending to become more common and such changes do not affect all species equally, since interacting communities involve species with diverse strategies, such response appearing species-specific. For Example, parasitoids of insect pests are linked to landscape profiles, making their response to spatial scale in relation to landscape important. Extensive monocrop industrial plantations have been raised since long, irreversibly altering the natural climax forest types. The human tropical forests by their potential for biomass generation and high genetic diversity and extreme vulnerability to irreversible damage are now engaging the serious attention of conservationists.

The present ``biotic crisis'' will alter not only biological diversity, but also the evolutionary processes by which diversity is generated. There will be several first order effects stemming from the biotic crisis : (i) a major extinction of species within the foreseeable future; (ii) a mega mass extinction of populations; (iii) alien invasions and other mixing of biotas; (iv) progressive depletion and homogenization of biotas with potential threshold effects ecosystem; (v) biotic impoverishment generally, possible including a decline of global biomass and (vi) gross reduction, if not virtual elimination of entire sectors of some biomass notably in tropical forests and wetlands, all of which have served as centers of diversification in the past. These first-order impacts will result in a series of further consequences viz : (i) fragmentation of species ranges with disruption of gene flow : (ii) decline in effective population sizes, with depletion of gene reservoirs / pools and (iii) biotic interchanges introducing species and even biotas into new areas, with multiple founder effects and novel competitive and other ecological interactions. These impacts, in turn, might disrupt food chains/webs, symbioses or biological associations.

Distinctive features of future evolution could include a homogenization of biotas, a proliferation of opportunistic species, a pest-and-weed ecology, an outburst of speciation among taxa that prosper in human dominated ecosystems, a decline of biodisparity, the depletion of ``evolutionary powerhouses'' in the tropics and unpredictable emergent novelties (Myers and Knoll, 2001). If we are concerned with avoiding the loss of particular functional groups, or with maximizing the potential source pool for evolutionary recovery, then biodisparity measures may provide a more appropriate assessment, beyond sheer numbers of taxa, of how priorities should be set (Jablonski, 1995). Against this background, it becomes imperative to give utmost priority to insect conservation because many insect species have immense breeding capabilities and rapid turnover rates. These attributes offer quick adaptability to environmental shifts whereupon genetic changes are passed along promptly. These attributes not only leave many insect species well suited to survice the environmental upheavals of human activities, but they offer exceptional scope for speciation in a comparatively short time span. An essential aspect of insect conservation in agroecosystems is the adoption of holistic ecological agricultural practices resting on the belief that nature encompasses everything else in a state of dynamic equilibrium and in the long run, nature attemps to resist any effort to change her, and that ultimately humans have no alternative, but to cooperate willingly or otherwise (Carson, 1963).

In short, conservation and sustainable development are closely interlinked and such an integrated approach demands satisfying basic human needs such as the ``maintenance and promotion of social, cultural and biological diversity and ecological integrity of the system" (Ramakrishnan, 2001). Climate change, rapid land degradation, desertification and biological invasion being responsible for global change, the role of biodiversity in copying up under the uncertainties of the environment becomes important, in particular species and subspecies-lerd interactions which are important for the functioning of ecosystem.

For choosing sites for conservation of fauna and flora, systematic information is necessary with accurate data. Systematic research plays a vital role in monitoring and managing our natural resources by providing accurate information on species level diagnosis, keys and distribution maps. Periodic monitoring of species is also needed. In today's scenario, phylogenetic studies serve as a foundation for an assessment of biological processes such as basic evolutionary trends, adaptation, speciation and extinction. Besides describing and sorting specing, phylogenetic trees through application of current molecular biological techniques like DNA and ribosomal RNA sequencing are to be constructed, reflecting taxonomy as determined by more traditional means of comparison.

REFERENCES

Ananthakrishnan, T.N. 1973. Mycophagous Tubulifera of India. Occ. Publ. Ent. Res. Inst. 114pp.

Ananthakrishnan, T.N. 1978. Microarthropods and soils Ecosystems. *Journal of Bombay National History Society,* 75 (3).

Ananthakrishnan, T.N. 1981. Thrips : in "*Vectors of Plant Pathogens*" (ed.) K.F. Harris and K. Maramorosch, Academic Press, N.Y. : 149-164.

Ananthakrishnan, T.N. 1982. Species dynamics in mycophagous thrips of the Oriental region *Ent. et. Gen.* 8 : 35-42.

Ananthakrishnan, T.N. 1984. *Bioecology of Thrips,* Indira Publ. House, Michigan, 233 pp.

Ananthakrishnan, T.N. 1986. *Bioresources Ecology,* Oxford & IBH, New Delhi 172 pp.

Ananthakrishnan, T.N. 1988. Dynamics of litter ecosystem in natural and intered forests. *Int. J. Ecol. Envir. Sci.,* 14 : 614-66.

Ananthakrishnan, T.N. 1992. *Dimension in Insect plant Interaction.* Oxford and IBH Publishing Co. Ltd., New Delhi, 184 pp.

Ananthakrishnan, T.N. 1995. *Biological Control of Social Forest and Plantation Crops* Oxford & IBH, New Delhi 225 pp.

Ananthakrishnan, T.N. and Gopichandran, R. 1993. *Chemical Ecology in Thrips host plant Interaction.* Oxford and IBH Publication, New Delhi, 125 pp.

Ananthakrishnan, T.N. and A. Raman 1989. *Thrips and Gall Dynamia* Leiden - E.J. Brill, 120 pp.

Ananthakrishnan, T.N. and A. Raman 1988. *Dynamics of Insect Plant Interaction.* Oxford-IBH, New Delhi, 222 pp.

Ananthakrishnan, T.N. and G. Suresh 1983. Paterns of fungal resource utilization and feeding range of some mycophagous Tubulifera (Insecta : Thysanoptera) *Proc. India Acad Sci.,* 92 (4) : 285-292.

Ananthakrishnan, T.N. and G. Suresh 1983. Association of some saprophagous thrips in forest communities with pathogenic fungi. *Indian J. Forestry* 6: 196-201.

Ananthakrishnan, T.N., B. Padmanabhan, K. Dhileepan and G. Suresh 1984. Ecological interactions species dynamics and reproductive biology of some mycophagous Thysanoptera *Occ. Publ. Ent. Res. Institute.* 5 : 37pp.

Ananthakrishnan, T.N. 1979. Biosystematics of Thysanoptera. *Ann. Rev. Entomol.* 24 : 159-183.

Ananthakrishnan, T.N. 1997. Insect biodiversity and conservation with reference to natural forest ecosystems. 301-308.

Ananthakrishnan, T.N. 1996. Forest litter Insect Conservation : Biology and Chemical Ecology Oxford & INH, New Delhi, 174 pp.

Anderson, A.N. 1997. Using ants as bioindicators : Multiscale issues in ant community ecology *Ecology and Society* 1 : 1-13.

Annadurai, R.S., Chandrasekhar, S.S. and Balu, A. 1988. Trophic structure and diversity in some litter inhabiting microarthropods of monoculture and natural forest ecosystems. *Proc. Indian Acad, Sci. (Anim. Sci.)* 97 : 301-308.

Appannah, S. 1990. Plant-pollinator interaction in Malaysian rainforests. In : K.S. Bawa and M. Hadley (eds.) *Reproductive Ecology of Tropical Rain Forest Plants.* Pantheon, Paris : 69-84.

Auerbach and Hendrix 1980. Insect-Fern Interaction *Ecol. Ent.* 5 : 99-104.

Balick, M.F., Furth, D.G. and Cooper Driver, G. 1978. Biological and Evolutionary aspects of arthropod predators on ferns. *Oecologia* (Berlin) 35 : 55-89.

Barrera, R. 1988. Multiple factors and their interactions on structuring the community of aquatic insects of tree holes. Ph.D. Thesis, The Pennsylvania State University, U.S.A.

Basset, Y. and M. Leponce 2005. IBISCA : Towards a census of canopy life ICAN, 12 : 1-11.

Bates, M. 1996. The *Forest and the Sea.* Randon House.

Batra, L.R. 1996. Ambrosia fungi : Extent of specificity to ambrosia beetles, *Science* 153 : 193-195.

Bezemer, T.M. and Jones, T.H. 1988. Plant herbivore interactions in elevated Co2 : quantitative analysis and guild effect *Oikos,* 82 : 212-222.

Bhatti, J.S., K. Vijayveer and B.K. Negi 1998. Discovery of the natural habitat of the aquatic thysanopteran, *Organothrips indicus* (Terebrantia : thysanoptera) in India and North America *Oriental Insects* 32 : 259-266.

Boulton, A.J., Findlay, S. Marmonier, P. Stanley, E.H. and Valet, H.m. 1998. The functional significance of the hyporheic zone in streams and rivers. Annu. Rev. ecol. Syst. 29 : 59-81.

Cardinale, B.J., Gelmann, E.R. and Palmer, M.A. 2004. Net spinning caddisflies as stream ecosystem engineers : the influence of *Hydropsyche* on benthic substrate stability. *Functional Ecology* 18 (3) : 381-387.

Chann, H.T. and Appannah, S. 1980. Reproductive biology of some Malaysian Dipterocarps 1 : Flowering biology. *Malaysian Forester,* 43 (2) : 132-143.

Chelliah, S. and M. Bharathi 1993. Insecticide induced resurgence of insects pests of crop plants. In : Pesticides : *Their Ecological impact in Developing Countries,* Commonwealth Publishers, New Delhi : 51-80.

Cheng, L. (ed.) 1976. Marine insects : North Holland Publishers, Amsterdam.

Cheng, L. 1985. Biology of *Halobates* (Heteroptera : gerridae) *Annual Review of Entomology* 30 : 111-135.

Connell, J.H. and Lowman 1980. Low diversity tropical and subtropical forests. *Amer Nat.* 134 : 89-119.

Corbet, S.A. 1995. Insects, plants and succession : advantages of long-term set aside. *Agr. Ecol. Envir.* 53 : 201-207.

Cuberson 1970. Chemosystematics and ecology of lichen forming fungi. *Ann. Rev. Ecol. Syst.* 1 ; 171-190.

Danks, H.V. and D.M. Rosenberg 1987.Aquatic insects of peatlands and marshes in Canada, *Mem. Ent. Soc. Can.* 140, 174pp.

Davis, G.M. 1995. Systematics and public health *Bioscience* 45 : 705-714.

Dery, M.S. and Davidar, P. 2003. Pollination system of trees in Kakachi, a mid elevation evergreen forest in western Ghats, India, *Amer, J. Bot.* 90 : 650-657.

Devy, M.S. and T. Ganesh 2003. Canopy Science and its relevance to India. *Curr. Sci.,* 85 (5) : 581-584.

Dhaliwal, G.S. and R. Arora 1996. *Principles of Insect Management :* National Agr. Tech. Infor. Centre. Ludhiana, 374 pp.

Dudgeon, D. 1999. Tropical Asian Streams-Zoobenthos : Ecology and Conservation. Hongkong Univ. Press, Hongkong, 828 pp.

Edgerly, J.S., A. Tadimalla and E.P. Dahlhoff 2005. Adaptation to thermal stress in lichen eating webspinners (Embioptera) : habitat choice, domicile construction and the probable role of heat shock proteins *Functional Biology* 19 : 255-262.

Ehrlich, P.R. and A.H. Ehrlich 1981. Extinction : The cause and consequences of disappearance of species. Randon House, N.Y.

Elton, C. 1933. The *Ecology of animals,* Methuen, London of disappearance of species. Randon House, N.Y.

Elton, C.S. 1958. The Ecology of Invasions by plants and animals. Chapman and Hall, London.

Erwin, T.L. 1982. Tropical Forests : their richness in Coleoptera and other arthropod species. *Coleopt Bull.* 36 : 74-82.

Erwin, T.L. 1983. Beetles and other insects of tropical forest canopies at Manaus, Brazil, Samples by insecticidal fogging. In: ``*Tropical Rain*

Forest : Ecology and Management : 59-75 S.L. Sutton, T.S. Whitman and A.C. Chadwich (eds.) Blackwell Publishers, Oxford.

Fahrig, L. 2003. Effects of habitat fragmentation in Biodiversity. *Ann. Rev. Ecol. Syst.* 34 : 487-515.

Fajer, E.D. 1989. How enriched Co2 enviroment may alter biotic systems even in the absence of climatic changes. *Conser. Biol.* 3 : 318-319.

Franklin, and Ehler 1978. Ecology of insects in urban environments. *Ann. Rev. Entomol.* 23 : 367-387.

Gibbs, J.P. and E.J. Stanton 2001. Habitat Fragmentation and arthropod community change : carrion beetles, phoretic mites and flies *Ecol. Appl.* 11 : 79-85.

Gillies, M.T. and Elouard, J.M. 1990. The mayfly-mussel association, anew example from the river Niger basin. IN : Campbell, I.C. (ed.) mayflies and Stoneflies. Kluwer Ac ademic Publishers, pp 298-297.

Gilbert, L.E. 1980. Food web organization and the conservation of Neotropical diversity : 11-34, in M.E. Soule and B.A. Wilcox (eds.) *Conservation Biology : An Evolutionary and Ecological Perspective,* Sinauer associates, Sunderland, MASS.

Ghosh, A.K. 1974. Fern infesting aphids (Insecta : Homoptera) in India. *J. Hort.* 31 : 104-109.

Graves, R.C. 1960. Ecological observations on the insects and other inhabitants of woody shelf fungi (Basidiomycetes : Polyporaceae) in Chicago area. *Ann. Ent. Soc. Amer.* 55 : 61-78.

Graves, R.C. and Graves, A.C.F. 1970. The insects and other inhabitants of shelf fungi on the Southern Blue Ridge region of Western N. Carolina IV. Thaysanoptera Ann. Entomol. Soc. America 63 : 96-98.

Heard, S.B. 1994a. Pitcher-plant midges and mosquitoes : a processing chain commensalisms. Ecology 75 : 1647-1660.

Heard, S.B. 1994b. Processing chain ecology : resource condition and interspecific interactions. Journal of Animal Ecology 63 : 451-464.

Hellawell, J.M. 1986. Biological Indicators of Freshwater : Pollution and Environmental Management Elsiveir, London.

Holter, P. 1979. Effects of dung beetles and earthworms on the disappearance of cattle dung *Oikos,* 32 : 393-402.

Huffaker, G. and Mathews, E.G. 1966. The natural history of dung beetles of the subfamily Scarabaeinae (Coleoptera : Scarabaeidae) *Folia Ent. Mex.* 12 :1-312.

Huxley, C. 1880-87. Symbiosis between ants and epiphytes. *Biol. Rev. 55* : 321-324.

Huxley, C.R. and F. Cutler 1991. *Ant-plant Interactions,* Oxford University Pres, Oxford.

Jagtap, A. and N.P. Singh 2002. Biodiversity of the Western Ghats of Maharashtra, Bishen Singh Mahendra Pal Singh, Dehra Dun.

Jaliano, S. a. 1989. Geographic variation in vulnerability to predation and starvation in larval treehole mosquitos. *Oikos* 56 : 99-108.

Jennersten, O. 1998. Pollination in *Dianthus dertoides* (Caryophyllaceae) Effect of habitat fragmentation on visitation and seed development. *Conserv. Biol.* 2 : 359-367.

Joseph, K.J. 1998. Biology and breeding behaviour of the elephant dung beetle, Heliocopris dominuns Bates (Coprinae : Scarabaeidae) Entomon 234 : 325-329.

Joseph, K.J. 2003. The life cycle, ecological role and biology of the immature stage of *Heliocopris dominus* Bates (Coleoptera : Scarabaeidae : Coprinae) Entomon, 28 (3) : 247-251.

Kadur, S. and K. Bawa 2005. *Sahyadris - India's Western Ghats* - a vanishing heritage.

Karr, J.R. 1991. Biological Integrity : A long neglected aspect of water resource management. Ecol. Appl. 1 : 66-84.

Kiihnelt, W. 1976. Soil Biology with special reference to the animal kingdom, Faber and Faber, London.

Kitching, R.L. 1971. An ecological study of water filled tree holes and their position in the woodland ecosystems. *J. Anim. Ecol.* 40 : 281-302.

Kitching, R.L. and Beaver, R.A. 1990. Patchiness and community structure. IN : Shorrocks, B. and Swingland, I.R. (eds.) Living in a patchy environment. Oxford University Press, Oxford U.K. pp. 147-176.

Kremen, C. 1992. Assessing the indicator properties of species assemblages for natural areas monitoring *Ecol. Appl.* 2 : 203-217.

Kremen, C., Colwell, R.K. Erwin, T.L. 1993. Terrestrial arthropod assemblages : Their use in conservation planning. *Conservat. Biol.* 7 : 796-808.

Kruess, K., and Tscharntke, T. 2000. Effects of habitat fragments on plant-insect communities. In : *Interchange of insects between agricultural and surrounding landscapes* (eds.) B. Ekbom, M.E. Irwin and Y. Robert, Kluwer Academic Publishers, London, 53-70.

Kumar and Prasad Entomon, 2 : 225-230.

Lakshminarayana, K.V., Yezdani, G.M. and C. Radhakrishnan 2002. Western Ghats, pp. 350-369. In : J.R.B. Alfred, A.K. Das and A.K. Sanyal, *Ecosystems in India.* Envis Centre, ZSI, Kolkata, 410 pp.

Lasky, K. 1997. The most beautiful roof in the world : Exploring the rainforest canopy Gulliver green. Harcour Brace & Co, San Diego.

Lawton, J.H. 2000. *Community Ecology in a changing world,* Excellence in Ecology series (ed.) Otto Kinne, Ecology Institute, Norbunde, Oldendorf, Germany, 227 pp.

Lines, J. 1995. Effects of climatiic and landuse changes on vectors of human diseases: 157-174. In : Harrington, R. and Stork, N.E. (eds.) 1995 *Insects in a changing Environment.* 17th Symposium Roy. Ent. Soc. London 535 pp.

Longino, J. and N. Nadkarni 1990. A comparison of ground and canopy leaf litter ants (Hymenoptera : Formicidae) in a Neotropical montane forest. *Psyche,* 97 : 81-94.

Lowman, M.D. and M. Moffat 1993. The Ecology of Tropical Rain Forest Canopies. *Tree* 8 (3) : 104-107.

Lowman, M.D. and N.M. Nadkarni 1995. *Forest Canopies* Academic Press San Diego.

Lowman, M.D. and P.K. Wittman 1996. Forest Canopies : methods, hypothesis and future direction. *Annu. Rev. Ecol. System* 27 : 55-81.

Mani, M.S. 1962. Introduction to high Altitude Entomology, Methuen, London.

Mathur, R.N. and Balwant Singh 1960. A list of insect pests of forest plants in India and adjacent countries. *Indian For.* 171 : 10-21.

Mckenzie, D.H., Hyatt, D.E. and V.J. Mcdonald 1995. *Ecological indicators* Chapman & Hall, London.

Mitchell, R. 1974. The evolution of thermophily in hot springs. Quarterly Review of Biology 49 : 229-242.

Moore, J.W. 2006. Animal Ecosystem Engineers in streams *Bioscience* 56 (3) : 237-246.

Moran, V.C. and T.R.E. Southwood, 1982. The guild composition of arthropod communities in trees. *J. Anim. Ecol.* 51: 289-306.

Mound, L.A. 1974. Spore feeding thrips (Phlaeothripidae) from leaf litter and dead wood in Australia. *Aust. J. Zool.* (Suppl.) 27 : 1-106.

Mound, L.A. 1974. The *Neurothrips* complex of spore feeding Thysanoptera (Phlaecothripidae : Idolothripinae) *Bull. Br. Mus. Nat. Hist.* 31 : 107-188.

Mound, L.A. 1993. The first thrips species inhabiting leaf domatia : *Domatiathrips cunninghami* Gen. et Sp. Nov. (Thysanoptera : Phlaeothripidae) J. New York Ent. Soc. 101 : 424-430.

Murphy, D.D., R.F. Noss and M.A. Sanjayan 2006. Urbanization as a major cause of biotic homogenization *Biol. Conserv.* 127 : 247-260.

Nadkarni, N.M. and J.T. Longino 1990. Invertebrates in canopy and ground organic matter in a Neotropical Montane Forest. *Biotropica* 22 (3) : 286-289.

Nash, T.M. 1996. *Lichen Biology*, Cambridge Univ. Press.

Noss, R.F. 1990. Indicators for monitoring biodiversity : A hierarchical approach. *Conserv. Biol.* 4 : 355-364.

Paoletti, M.R., Taylor, B., Stinner, D., and Benzing, D. 1991. Diversity of soil fauna in the canopy and forest floor of a Veneuzuelan Cloud Forest J. Trop. Ecol. 7 : 373-383.

Paradise, C.J. and Dunson, W.A. 1997. Insect species interactions and resource effects in tree holes : Are helodid beetles botton-up facilitators of midge populations? Oecologia 109 : 3036-312.

Paulson, D. 2005. Dragonflies in the canopy, Global Canopy Programme 12 (1) : 7-9.

Paulson, D. 2005. Dragonflies in the canopy. What is up? Newsletter of Invertegrate canopy Network, 12 (1) : 7-8.

Peter Farb 1990. Ecology. Time & Life books, N.Y. 192 pp.

Poykko, H. 2005. Host range of lichenivorous moths with special reference to nutritional quality and chemical defence in lichens, *kirjasto* : 1-2.

Priesner, H. and A.R. Seshadri, 1952. Some new thysanoptera from South India - I. *Indian J. Agric. Sci.,* 22: 405-411.

Rao, M.V., Gupta, G.K. and Dubey, P.S. 1990. Effects of relatively low ambient air pollution on total sucking insect population on *Sorghum vulgare Trop. Ecol. 31* : 66-72.

Ramakrishnan, P.S. 2001. Ecology and sustainable development. National Book Trust, India, 198pp.

Ramakrishnan, P.S. and O.P. Toky, 1981. Soil nutrient status of hill agrosystems and recovery patterns after slash and burn agriculture in N.E. India Plant and Soil, 60 : 41-64.

Reynolds, B.C. and D.A. Crossley 1997. Spatial variation in herbivores of forest canopy arthropods along an elevation gradient. *Ecol. Entomol.* 25 : 348-356.

Rinker, H.B. and M.D. Lowman 2001. Literature Review : Canopy herbivores and soil Ecology : The top down impact on forest processes. *Selbyana* 22 (2) : 225-231.

Rukke, E. 2000. Effects of habitat fragmentation : increased isolation and reduced habitat size reduces the incidence of dead wood fungi beetles in a fragmented landscape *Ecography* 23 : 492-502.

Samways, M.J. 1995. Southern hemisphere Insects : Their variety and environmental processes upon them. In : Harrington, R. and Stork, N.E. (eds.) *Insects in a changing Environment.* 17th Symp. R. Ent. Soc. London Academic Press, London, 535 pp.

Schowalter, T.D., Hargrove, W.W. and D.A. Crossby 1986. Herbivory in forested ecosystems. *Ann. Rev. Entomol.* 31:177-196.

Scott, Miller, L.M.E. Soute, and D.F. Doak 1993. The Keystone concept in Ecology and Conservation Bioscience 43 (4) : 219-223.

Seasted, T.R. 1984. The role of microarthropods in decomposition and minerlization process. *Ann. Rev. Entomol.* 29 : 25-49.

Seasted, T.R. and D.A. Crossley 1984. The influence of arthropods on Ecosystem *Bioscience* 34 : 157-161.

Seifert, R.P. and Seifert, F.N. 1976. A community matgrix analysis of Heliconia insect communities. American Naturalist 110 : 461-483.

Seifert, R.P. and Seifert, F.N. 1979. A Heliconia insect community ion a Venezuelan cloud forest Ecology 60 : 462-467.

Shivshanker, T., A.R.V. and G.K. Veeresh 1995. Ponerine ants an probable predators of termites under forest ecosystems. *In : Biological control of Social Forest and Plantation Crop Insects,* (ed.) Oxford & IBH, New Delhi, 136 - 151.

Singh, J.S. and S.R. Gupta 1977. Plant decomposition and soil respiration in terrestrial ecosystems. *Bot. Rev.* 43 : 449-528.

Sivaramakrishnan, K.G., Hannaford, M.J. and Resh, V.H. 1996. Biological assessment of Kaveri river catchment, South India, Using benthic macro invertebrates : Applicability of water quality monitoring approaches developed in other countries. *Int. J. Ecol. Envir. Sci.* 22 : 113-132.

Sivaramakrishnan, K.G., Venkataraman, K. Sridhar, s. and Marimuthu, M. 1995. Spatial patterns of benthic macroinvertebrate distribution along river Kaveri and its tributaries (India). *Int. J. Ecol. Envir. Sci.* 1 : 141-161.

Smith, A.J.E. 1982. *Bryophyte ecology,* Chapman & Hall.

Stephen, C.J., Schellhoven, N.a., Wood, G.M., and Austin, A.D. 2006. Parasite-wasp assemblage associated with natural and weed plant species in an agricultural landscape *Aust. J. Ecol.* 45 : 176-184.

Stevenson, B.G. and Dindal, D.L. 1987. Functional Ecology of coprophagous beetles *Pedobiologia* 30 : 285-298.

Stork, N.E. and Blackburn, J.M. 1993. Abundance, Body size and Biomass of arthropods in Tropical Forests, *Oikos* 67 : 483-489.

Stork, N.E., J.A. Adis and R.K. Didham 1997. *Canopy Arthropods,* Chapman & Hall, London.

Subramanian, K.A., Sivaramakrishnan, K.G. and Madhav Gadgil 2005. Impact of riparian landuse on stream insects of Kudremukh National Park, Karnataka State, India, J. Insect Sci , 5 : 49.

Subramanian, K.A. and Sivaramakrishnan, K.G. A new record of the mayfly-mussel association from Tunga River in Karnataka state, India, with a description of a new species of Symbiocloeon (Ephemeroptera : Baetidae) (manuscript submitted to Zootaxa).

Suthurst, R.W., Maywald, G.F. and Skaratt 1995. Predicting insect distribution in a changing climate : 59-91. In : Harrington, R. and Stork, N.E. (eds.) *Insects in a changing environment* 17th Symp. R. Ent. Soc. London Academic Press, London, 535pp.

Stork, N. 1991. The composition of arthropod fauna of Borneo lowland rainforest. *J. Trop. Ecol.* 7 : 161-180.

Terborgh, J. 1986. Keystone plant resources in the tropical forest. In : M.E. Soule (ed.) *Conservation Biology : the science of scarcity and diversity.* Sinauer Associates, Sunderland, MASS.

Theinmann, A. 1954. *Chironomus* Leber Verbreitang wirtschafte biche Bedeutung der Chironomidae. *Die Binnengewesser* 20 : 1-834.

Thornton, I.W.B. 1985. The geographical and ecological distribution of arboreal Psocoptera. *Ann. Rev. Entomol.* 30 : 175-196.

Thorpe, W.H. 1930. The biology of the petroleum fly (*Psilopa petroleii* Coq.) Transactions of the Entomological Society of London 78 : 331-343.

Turner, M.G. 1989. Landscape Ecology : The effect of pattern on process, *Ann. Rev. Entomol.* 20 : 171-197.

Uri Gerson 1982. *Bryophytes and Invertebrates.* In : *Bryophyte Ecology* (ed.) A.J.E. Smith Chapman & Hall, London.

Vanderwoude, C., A.N. Anderson and P.N. House 1997. Ant communities as bioindicators in relation to fire management of spotted gum (*Eucalyptus maculates)* forests in South East Queensland. *Mem. Mus. Victoria* 56 : 671-675.

Veeresh, G.K., T.M. Mustak, Ali and A.R.V. Kumar 1995. Ants as balancer in nature for pest problems. In : *Biological control of Social Forest and Plantation Crops insects.* Ed. T.N. Ananthakrishnan Oxford & IBH, New Delhi : 121-136.

Waldbauer, G. 2006. A walk around the Pond : Insects in an over the water. Harward University Press, Cambridge Massachusetts 286 pp.

Walker, I.R., Smith, J.P., Engstrom, D.R. and Birks, H.J.B. 1991. An assessment of chironomids as qualitative indicators of past climate change. *Can. J. Fish Aquat Sci. 48* : 975-987.

Watt, A.D., Whittaker, J.B., Docherty, M., Brooks, G., Lindsay, E. and Salt, D.J. 1995. The impact of elevated Co^2 in Insect herbivores. In : Harrington, R. and Stark, N.E. 1995. *Insects in a changing environment* 17th Symp. Roy. Ent. Soc. London Academic Press, London 535pp.

Weber, N.A. 1966. Fungus growing ants, Science 153 : 587-604.

Wilcore, A.S., Mclellan, C.H. and Dobson, A.P. 1986. Habitat Fragmentation in the Temperate Zone. In : *Conservation Biology* (ed.) M.E. Soule : 237-256. Sunderland, M.A. Sinauer.

Wilcove, D.S., Mclellan, D.S. and A.P. Dobson 1981. Habitat fragmentation in the temperate Zone. In : *Conservation Biology* (ed.) M.E. Soule : 23-56. Sinauer Associates, Sunderland.

William, D.D. 1987. *The Ecology of Temporary Waters,* Croom Helm, London.

Williams, D.D. 1974. The hyprorheic Zone as a habitat for Aquatic Insects and Associated Arthropods. IN : Resh, V.H. and Rosenberg, D.M. (eds.) The Ecology of Aquatic Insects, Praeger Publ. Co., New York 430-455.

Williams, D.D. and Williams, N.E. 1998. Aquatic insects in an estuarine environment : densities, distribution and salinity tolerance Freshwater Biology 30 : 411-421.

Wilson, E.O. 2003. *The future of life.* Vintage Books, N.Y. 229 pp.

Wolda, H. 1979. Abundance and diversity of Homoptera in the canopy of a tropical forest. *Ecol. Ent.* 4 : 181-190.

Zanetell, B.A. and Peckarsky, B. 1996. Stoneflies as ecological engineers - hundry predators reduce fine sediments in stream beds. *Freshwater Biology* 36 (3) : 569-577.

SUBJECT INDEX